MW00339725

Guidelines for
Auditing Process Safety
Management Systems

Publications Available from the
CENTER FOR CHEMICAL PROCESS SAFETY
of the
AMERICAN INSTITUTE OF CHEMICAL ENGINEERS

Guidelines for Auditing Process Safety Management Systems

Guidelines for Investigating Chemical Process Incidents

Guidelines for Hazard Evaluation Procedures, Second Edition with Worked Examples

Plant Guidelines for Technical Management of Chemical Process Safety

Guidelines for Technical Management of Chemical Process Safety

Guidelines for Chemical Process Quantitative Risk Analysis

Guidelines for Process Equipment Reliability Data, with Data Tables

Guidelines for Vapor Release Mitigation

Guidelines for Safe Storage and Handling of High Toxic Hazard Materials

Guidelines for Use of Vapor Cloud Dispersion Models

Safety, Health, and Loss prevention in Chemical Processes: Problems for Undergraduate Engineering Curricula

Safety, Health, and Loss prevention in Chemical Processes: Problems for Undergraduate Engineering Curricula—Instructor's Guide

Workbook of Test Cases for Vapor Cloud Source Dispersion Models

Proceedings of the International Conference on Hazard Identification and Risk Analysis, Human Factors, and Human Reliability in Process Safety, 1992

Proceedings of the International Conference/Workshop on Modeling and Mitigating the Consequences of Accidental Releases of Hazardous Materials, 1991.

Proceedings of the International Symposium on Runaway Reactions, 1989

Proceedings of the International Conference on Vapor Cloud Modeling, 1987

Proceedings of the International Symposium on Preventing Major Chemical Accidents, 1987

1991 CCPS/AIChE Directory of Chemical Process Safety Services

Audiotapes and Materials from Workshops at the International Conference on Chemical Process Safety Management, 1991

Electronic Chemical Process Quantitative Risk Analysis Bibliography

Guidelines for Auditing Process Safety Management Systems

WILEY-INTERSCIENCE

A JOHN WILEY & SONS, INC., PUBLICATION

CENTER FOR CHEMICAL PROCESS SAFETY

of the

AMERICAN INSTITUTE OF CHEMICAL ENGINEERS

345 East 47th Street, New York, New York 10017

Copyright © 1993
American Institute of Chemical Engineers
345 East 47th Street
New York, New York 10017

No part of this publication may be reproduced, stored in a retrieval system or transmitted in any form or by any means, electronic, mechanical, photocopying, recording, scanning or otherwise, except as permitted under Sections 107 or 108 of the 1976 United States Copyright Act, without either the prior written permission of the Publisher, or authorization through payment of the appropriate per-copy fee to the Copyright Clearance Center, 222 Rosewood Drive, Danvers, MA 01923, (978) 750-8400, fax (978) 646-8600. Requests to the Publisher for permission should be addressed to the Permissions Department, John Wiley & Sons, Inc., 111 River Street, Hoboken, NJ 07030, (201) 748-6011, fax (201) 748-6008.

To order books or for customer service please, call 1(800)-CALL-WILEY (225-5945).

Library of Congress Cataloging-in Publication Data
Guidelines for auditing process safety management systems / Center
   for Chemical Process Safety of the American Institute of Chemical
   Engineers.
       p.   cm.
   Includes bibliographical references and index.
   ISBN 0-8169-0556-8
   1. Chemical plants—Safety measures.   I. American Institute of
Chemical Engineers. Center for Chemical Process Safety.
TP149.G835   1993
860' .2804—dc20   92-40116
                                                  CIP

Second printing 1993

This book is available at a special discount when ordered in bulk quantities. For information, contact the Center for Chemical Process Safety at the address shown above.

It is sincerely hoped that the information presented in this volume will lead to an even more impressive safety record for the entire industry; however, neither the American Institute of Chemical Engineers, its consultants, CCPS and/or its sponsors, its subcommittee members, their employers, nor their employers' officers and directors warrant or represent, expressly or implied, the correctness or accuracy of the content of the information presented in this conference, nor can they accept liability or responsibility whatsoever for the consequences of its use or misuse by anyone.

Contents

Chapter 3   Accountability and Responsibility

Chapter 4   Process Safety Knowledge

Chapter 11   Training and Performance

Chapter 12   Emergency Response Planning

List of Figures and Tables

Acronyms

AIChE	American Institute of Chemical Engineers
AIChE-DIERS	American Institute of Chemical Engineers—Design Institute for Emergency Relief Systems
ANSI	American National Standards Institute
API	American Petroleum Institute
ASME	American Society of Mechanical Engineers
CAD	Computer Aided Design
CCPS	Center for Chemical Process Safety
CMA	Chemical Manufacturers Association
EHSRMA	Extremely Hazardous Substances Risk Management Act (DE)
EPA	Environmental Protection Agency
FMEA	Failure Modes and Effects Analysis
HAZOP	Hazard and Operability Analysis
HVAC	Heating, Ventilating and Air Conditioning
MSDS	Material Safety Data Sheet
NDE	Non-Destructive Examination
OSHA	Occupational Safety and Health Administration
PFD	Process Flow Diagram
P&ID	Piping and Instrument Diagram
RCRA	Resource Conservation and Recovery Act
RMPP	Risk Management and Prevention Program (California)
SARA	Superfund Amendments and Reauthorization Act
SOP	Standard Operating Procedure
TCP	Toxic Catastrophe Prevention Act (New Jersey)
UPS	Uninterruptable Power Supply

Glossary

Accident: An incident limited to a single injury and/or minor property damage.

Accountability: The obligation to explain and answer for one's actions that are related to expectations, objectives, and goals. Because it is associated with positive and negative rewards for actions taken, accountability gives "teeth" to the roles and responsibilities assigned through the management system. Accordingly, it is a powerful element of an effective process safety management system.

Action plan: A project schedule for the follow-up activity, and a management control document which can be used to monitor the status of corrective action.

Administrative control: Procedures that will hold human and/or equipment performance within established limits.

Audit: A systematic, independent review to verify conformance with established guidelines or standards. It employs a well-defined review process to ensure consistency, and to allow the auditor to reach defensible conclusions.

Checklist (traditional): A detailed list of desired system attributes or steps for a system or operator to perform. Usually written from experience and used to assess the acceptability or status of the system or operation compared to established norms.

Confirmation: A special audit term referring to the substantiation of the existence or condition of something. A confirmation often takes the form of a written request and acknowledgement from independent third parties, but it may also be obtained orally or through observation.

Consequence: The direct, undesirable result of an accident sequence usually involving a fire, explosion, or release of toxic material. Consequence descriptions may be qualitative or quantitative estimates of the effects of an accident in terms of factors such as health impacts, economic loss, and environmental damage.

Consistency: Continued uniformity, during a period or from one period to another.

Determine: To conclude; to reach an opinion consequent to the observation of the fit of sample data within the limit, range, or area associated with substantial conformance, accuracy, or other predetermined standard; to obtain firsthand knowledge of.

Evaluate: To reach a conclusion as to significance, worth, effectiveness or usefulness.

Exception: A finding which is a deviation from a standard.

Failure Modes and Effects Analysis (FMEA): A systematic, tabular method for evaluating and documenting the causes and effects of known types of component failures.

Fault tree: A logic model that graphically portrays the combinations of failures that can lead to a specific main failure or accident of interest.

Finding: A conclusion, positive or negative, reached by the audit team based on data collected and analyzed during the audit. Negative findings are typically called exceptions.

Frequency: The number of occurrences per unit time at which observed events occur or are predicted to occur.

Hazard: An inherent physical or chemical characteristic that has the potential for causing harm to people, property, or the environment. In this document it is the combination of a hazardous material, an operating environment, and certain unplanned events that could result in an accident.

Hazard analysis: See hazard evaluation.

Hazard evaluation: The analysis of hazardous situations associated with a process or activity, using qualitative techniques to identify weaknesses in design and operation.

Hazard identification: The recognition of material, system, process, and plant characteristics that can produce undesirable consequences through the occurrence of an accident.

Hazard and Operability (HAZOP) Analysis: A systematic method in which process hazards and potential operating problems are identified using a series of guide words to investigate process deviations.

Human factors: A discipline concerned with designing machines, operations, and work environments to match human capabilities, limitations, and needs. Among human factors specialists, this general term includes any technical work (e.g., engineering, procedure writing, worker training, worker selection) related to the person in operator-machine systems.

Incident: An unplanned event with the potential for undesirable consequences.

Inspection: The process of physically examining a facility.

Internal controls: The various engineering and managerial means, both formal and informal, established within an organization to help the organization direct and regulate its activities in order to achieve desired results; also refers to the general methodology by which specific management processes are carried on within an organization.

Internal controls questionnaire: A questionnaire specifically designed to assist in efficient collection of general or specific background information about the facility's internal management systems and procedures.

Interviewing: Questioning, both formally and informally, facility personnel or other individuals in order to obtain an understanding of the plant's operations and performance.

Likelihood: The expected frequency of an event's occurrence.

Major accident: An incident involving multiple injuries, a fatality, and/or extensive property damage.

Near-miss: An incident that has the potential for injury and/or property damage.

Objectivity: Freedom from bias.

Observation: The noting and recording of information to support findings.

Operator: An individual responsible for monitoring, controlling, and performing other tasks as necessary to accomplish the productive activities of a system. Often used in a generic sense to include people who perform various tasks (e.g., reading, calibration, maintenance).

Process safety: The protection of people and property from episodic and catastrophic incidents that may result from unplanned or unexpected deviations in process conditions.

Process safety auditing: A formal review that identifies process hazards relative to established standards; for example, examining plant and equipment, often using a checklist or other audit guide.

Process safety management: The application of management systems to the identification, understanding, and control of process hazards to prevent process-related incidents and injuries.

Process safety management systems: Comprehensive sets of policies, procedures, and practices designed to ensure that barriers to episodic incidents are in place, in use, and effective.

Process safety management systems auditing: The systematic review of process safety management systems, used to verify the suitability of these systems and their effective, consistent implementation.

Protocol: A document which organizes audit procedures into a general sequence of audit steps and describes such steps in terms that specify the actions to be taken by the auditor.

Review: To study critically an operation, procedure, condition, event, or series of transactions.

Risk: The combination of the expected frequency (events/year) and consequence (effects/event) of a single accident or a group of accidents.

Risk assessment: The process by which the results of a risk analysis (i.e., risk estimates) are used to make decisions, either through relative ranking of risk reduction strategies or through comparison with risk targets.

Risk management: The systematic application of management policies, procedures, and practices to the tasks of analyzing, assessing, and controlling risk in order to protect employees, the general public, the environment, and company assets.

Risk measures: Ways of combining and expressing information on likelihood with the magnitude of loss or injury (e.g., risk indices, individual risk measures, and societal risk measures).

Root causes: Management system failures, such as faulty design or inadequate training, that led to an unsafe act or condition that resulted in an incident; underlying cause. If the root causes were removed, the particular incident would not have occurred.

Safety review: An inspection of a plant or process unit, drawings, procedures, emergency plans, and/or management systems, etc., usually by a team and usually problem-solving in nature. (See "Audit" for contrast).

Sample/sampling: Selecting a portion of a group of data in order to determine the accuracy or propriety or other characteristics of the whole body of data.

Standard: Any established measure of extent, quantity, quality or value. Any type, model or example for comparison; a criterion of excellence.

Task analysis: A human error analysis method that requires breaking down a procedure or overall task into unit tasks and combining this information in the form of event trees. It involves determining the detailed performance required of people and equipment and determining the effects of environmental conditions, malfunctions, and other unexpected events on both.

Toll processors: Contract chemical processors who produce material for a company.

Topical outline: A short summary or list of the major topics to be covered during the audit. As a simple list of key subjects, the topical outline relies to a great extent on the experience and judgment of the auditor.

Verification: A wide variety of activities that can be employed to increase confidence in the audit data, including: evaluating the application of, and adherence to, laws, regulations, policies and procedures, standards and management directives; certifying the validity of data and reports; and evaluating the effectiveness of management systems.

Verify: To confirm the truth, accuracy, or correctness of, by competent examination; to substantiate.

What-if analysis: A brainstorming approach in which a group of experienced people familiar with the subject process ask questions or voice concerns about possible undesired events.

Working papers: Field notes used in preparation of the final report documenting work performed, techniques used and conclusions reached while conducting the audit.

Preface

The American Institute of Chemical Engineers (AIChE) has a 30 year history of involvement with process safety and loss control issues in the chemical, petrochemical, and hydrocarbon process industries. AIChE publications and symposia are information resources for the chemical engineering profession on the causes of process incidents and means of preventing their occurrences or mitigating their consequences.

The Center for Chemical Process Safety (CCPS), a directorate of AIChE, was established in 1985 to develop and disseminate technical information for use in the prevention of major chemical process incidents. With the support and direction of the CCPS Advisory and Managing Boards, a multifaceted program was established to address the need for process safety management systems in industry to reduce potential exposures to the public and the environment. Over 80 corporations from all segments of the process industries provide the funding and professional experience for the Center's activities.

In 1989, CCPS published the *Guidelines for Technical Management of Chemical Process Safety*, which developed a model for a safety management system characterized by twelve distinct and essential elements. The Foreword to that project states:

"For the first time, all the essential elements and components of a model of a technical management program in chemical process safety have been assembled in one document. We believe these *Guidelines* provide the umbrella under which all other CCPS Technical Guidelines will be promulgated."

One of the twelve elements in the model developed in the *Guidelines for Technical Management of Chemical Process Safety* is *Audits and Corrective Actions*, which is the subject of this book. The intent of this book is to provide fundamental information for developing an audit program to help ensure that the elements of a process safety management system are in place and functioning. As such, it provides guidance for auditing the other elements in a comprehensive process safety management systems auditing program. A sound auditing program for process safety management systems can reinforce the effectiveness of the entire process safety system.

The first three chapters in the *Guidelines* provide guidance pertaining to the management of the audit program:

Chapter 1—Management of Process Safety Management Systems Audits
Discusses a number of choices on issues relating to the design of an audit program for process safety management systems. These issues include audit scope, frequency,

staffing, reporting, follow-up, and quality assurance. It emphasizes the importance of clearly defining the program objectives and developing a consistent approach.

Chapter 2—Audit Techniques

Describes various audit activities and tools, including preparation, audit guides, methods for gathering data, evaluating field work, recordkeeping, and follow-up. Provides examples of audit guides and interviewing techniques.

Chapter 3—Accountability and Responsibility

Discusses the indicators of accountability and responsibility which should be considered in a process safety management systems audit. As accountability and responsibility are principles rather than activities, they are difficult to audit; therefore, the auditor needs to identify specific indicators.

The remaining nine chapters discuss the auditing of the elements in a comprehensive process safety management system. These include

This book contains information useful to both experienced auditors and those developing an audit program. It presents state-of-the-art techniques and methods that should be useful to auditors in the process industries. Consistent application of these techniques and analysis of results will contribute to continuous improvement in process safety management.

Acknowledgments

The American Institute of Chemical Engineers and the Center for Chemical Process Safety thanks all of the members of the Process Safety Management Audit Subcommittee for their dedicated efforts and technical contributions to the preparation of the Guidelines. CCPS also expresses appreciation to the members of the Technical Steering Committee for their advice and support.

The Chair of the Process Safety Management Audit Subcommittee was Marvin F. Specht of Hercules Incorporated. The Subcommittee members were William S. Turetsky, ISP/GAF Corporation; Donald C. Clagett, General Electric Company; Herm Waltemate, BF Goodrich; Dale M. Shapiro, Hoechst Celanese; Albert Kover, The Lubrizol Corporation; Dale Schillinger, Mallinckrodt Specialty Chemicals Co.; Robert S. Cutro, Merck & Co. Inc.; K. Gerry Phillips, Novacor Chemicals Ltd; David G. Kehn, Occidental Chemical Corporation; and Stanley E. Anderson, Rohm and Haas Texas Incorporated. Ray E. Witter was the CCPS staff liaison and was responsible for the overall administration and coordination of the project.

The members of the Process Safety Management Audit Subcommittee also wish to thank their employers for providing time to participate in this project and to the many sponsors whose findings made this project possible.

Arthur D. Little, Inc., Cambridge, Massachusetts, was the contractor for this project. Henry Ozog was Arthur D. Little's Project Director. R. Scott Stricoff served as Officer-in-Charge. The principal authors were: P.J. Bellomo, Lisa M. Bendixen, Maryanne DiBerto, Paul M. Dixon, Frederick T. Dyke, Gilbert S. Hedstrom, Marian H. Long, Henry Ozog, Christine A. Sabatke, R. Peter Stickles, and R. Scott Stricoff. Dana Pierce was the Technical Editor. Lucie Leveille and Tracey Martensen provided secretarial and graphics support.

CCPS also gratefully acknowledges the comments and suggestions submitted by the following peer reviewers: Prabir K. Basu, GD Searle; Brian D. Berkey, Hercules, Inc.; L.O. Bowler, General Electric Company; Ronald Bussey, Merck & Co., Inc.; George G. Buxton, Occidental Petroleum Corporation; Daniel A. Crowl, Wayne State University; Charles Dancer, Allied Signal; Art Dowell, Rohm & Haas Texas; Denny Dowell, General Electric Company; Thomas G. Fisher, The Lubrizol Corporation; Peter D. Fletcher, Badger Design & Constructors; T.O. Gibson, Dow Chemical Company; Jay E. Giffin, Union Carbide; Robert J. Grahek, BF Goodrich; John T. Higgins, Dow Corning; Robert E. Holm, Occidental Chemical Company; Peter Hughes, Novacor Chemicals Ltd.; Dave Mack, Novacor Chemicals Ltd.; Michael T. McHale, Air Products; Gregory C. Noll, Hildebrand and Noll Associates, Inc.; Homer

Richardson, Consultant; Gary Van Sciver, Rohm and Haas; Robert C. Wade, Amoco Oil Company; Jan Windhorst, Novacor Chemicals Ltd; and Jack F. Yablonsky, General Electric Company. Their insight and thoughtful comments helped ensure a balanced perspective for the *Guidelines*.

Introduction

An audit is a fundamental part of an effective process safety management program because its purpose is to verify that systems to manage process safety are in place and functioning effectively. The audit element also needs to have a management system in place to ensure that it functions effectively—particularly the follow-up on action items. Equally important is that auditors have the proper skills and tools to audit effectively.

A comprehensive audit of process safety management systems can be accomplished using different approaches. This book provides alternatives for developing audit programs to meet the needs of a variety of companies from small businesses to international corporations. This book also addresses some basic skills, techniques, and tools that are fundamental to auditing, and some characteristics of good process safety management systems that an auditor should be looking for in facility programs.

The information that must be gathered and evaluated during an audit will vary considerably from facility to facility and process to process. Information that an auditor is looking for may reside in more than one location or may not be documented. Therefore, this book provides guidance on information that an auditor may need to review, and what to look for.

Regardless of the approach and techniques used to conduct process safety management systems audits, the most important aspects are that the audits be objective, be systematic, and be done periodically.

1

Management of Process Safety Management Systems Audits

1.1 Overview

In its earlier publications, the American Institute of Chemical Engineers' Center for Chemical Process Safety (CCPS) defined 12 *elements* of process safety management. Management systems that address each of these 12 elements are needed in a comprehensive process safety management program. The 12 elements defined by CCPS (listed in Table 1-1) represent one way of expressing concepts that are fundamentally similar to those expressed in other recent work on process safety management.

Auditing is one of the 12 CCPS process safety management elements. It is a critical element in that it contributes to management control of the other elements. A sound process safety management auditing program will improve the effectiveness of an entire process safety program.

In discussing process safety management auditing, some confusion over terminology may arise. "Auditing" is used in various contexts to describe many different types of activities, and "process safety management" is a term that is still relatively new. In this book, the following definitions are used:

- An *audit* is a systematic, independent review to verify conformance with established guidelines or standards. It employs a well-defined review process to ensure consistency, and to allow the auditor to reach defensible conclusions.
- An *inspection* is the process of physically examining a facility.

Less formal reviews, which may combine aspects of inspections and audits, are guided by the judgment, experience, and inclination of the reviewer, often without a well-defined review procedure or process. Such a review often has a broader scope than an inspection, but it does not have the consistency and rigor of an audit.

Other definitions germane to this document are presented below:

- *Process safety* refers to the protection of people and property from episodic and catastrophic incidents that may result from unplanned or unexpected deviations in process conditions. (This is an ideal condition toward which one strives. However, the handling, use, storage, and processing of materials with inherent hazardous properties can never be absolutely free from risk.)

1

TABLE 1-1
Twelve Elements of Chemical Process Safety Management

Accountability: Objectives and Goals
Process Knowledge and Documentation
Capital Project Review and Design Procedures
Process Risk Management
Management of Change
Process and Equipment Integrity
Incident Investigation
Training and Performance
Human Factors
Standards, Codes, and Laws
Audits and Corrective Actions
Enhancement of Process Safety Knowledge

Source: CCPS (1989), *Guidelines for Technical Management of Chemical Process Safety.*

- *Process safety management* is the application of management systems to the identification, understanding, and control of process hazards to prevent process-related incidents and injuries.
- *Process safety management systems* (PSM systems) are comprehensive sets of policies, procedures, and practices designed to ensure that barriers to episodic incidents are in place, in use, and effective.
- *Process safety auditing* is a formal review that identifies process hazards relative to established standards; for example, examining plant and equipment, often using a checklist or other audit guide.
- *PSM systems auditing* is the systematic review of PSM systems, used to verify the suitability of these systems and their effective, consistent implementation.

Process safety auditing is fundamentally different from PSM systems auditing. Where the focus of the former type of activity falls on the identification and evaluation of specific hazards, the focus of the latter falls on assessment and verification of the management systems that ensure on-going hazard control.

This important distinction is clearly illustrated with an example:

One can inspect the piping and equipment for a process and identify the absence of a required pressure relief device in the system.

One can review the management systems in place to ensure that pressure relief devices have been designed, installed, operated, and maintained in accordance with company standards.

The first of these reviews addresses a particular hazard found at a specific time. It could lead to correction of that hazard without addressing the underlying reason why that hazardous condition came to exist. Alternatively, the second of these reviews addresses the management system in place to preclude the creation of hazards. Detection of a breakdown in the *system* could lead to precluding the creation of hazards in the future.

Both types of reviews are important in enhancing process safety. This book addresses the *PSM systems audit.*

PSM systems audits are intended to determine whether management systems are in place and functioning properly to ensure operating facilities and process units facilities and process units have been designed, constructed, operated, and maintained such that the safety and health of employees, customers, communities, and the environment are being properly protected. These audits are an important control mechanism within the overall management of process safety. In addition, these audits can provide other benefits such as improved operability and increased safety awareness.

The criteria used during such audits may be limited to the requirements of specific laws and regulations, or they may be broadened to include company policies and standards, or the guidelines of groups such as CCPS. Each company should decide on appropriate audit criteria during the design of its audit program.

A PSM systems audit involves examination of management system design, followed by evaluation of management system implementation. The design of the management system must be understood and then evaluated to determine if the system, when functioning as intended, will meet the applicable criteria. Then the auditor must evaluate the quality and degree of implementation since a sound system design may not be backed up by consistent, thorough implementation.

This chapter discusses the issues associated with the design and management of a PSM systems auditing program. Specifically, the issues of audit scope, frequency, staffing, reporting, follow-up, and quality assurance are discussed.

1.2 Audit Program Scope

The scope of an audit program refers to the facilities and units to be covered, the subject areas to be addressed, and the criteria against which the audit is to be conducted. Process safety management system audits can vary considerably in scope.

It is important that the scope of the audit program be clearly defined. Failure to do so can lead to misunderstandings among the groups being audited, the auditors, and the management recipients of the audit. Failure to define the scope of an audit program can also lead to inconsistent and inaccurate audit results, to findings being missed, or to the inclusion of inappropriate observations in audit reports.

When defining the scope of a PSM systems audit program, a number of factors should be considered. Among these are

- Company policies
- Regulatory requirements
- Resource limitations
- Time available
- Nature of operations and risks
- Other management control mechanisms

Where company policies related to process safety exist, these must be considered in designing the audit program. For example, company policy may dictate audit frequency. Where regulations call for process safety management system audits, these requirements will influence the design of the program.

An audit program may have its scope defined as including all facilities, or only manufacturing facilities, or only facilities handling certain hazardous materials. It may cover only wholly owned facilities, or it may cover joint ventures and partnerships, or it may extend to the contract chemical processors (often known as "toll processors") who produce material for the firm. Among the parameters considered in defining the scope of the audit program are

- Type of facility (manufacturing, terminals, etc.)
- Ownership (wholly owned, joint ventures, etc.)
- Geographical location
- Site coverage (all units versus selected units)
- Program content (all process safety management elements versus selected elements)

A practical consideration in defining the scope of an audit program is the availability of resources. The scope should be considered, together with the issues of audit frequency (discussed in section 1.3), and resources available to develop a program that addresses the range of operations and the risks.

The time available for the audit should also be considered in designing the audit program scope. It is better to perform a thorough audit with a narrower scope than to perform a hurried, incomplete audit with a broader scope. The latter exercise often results in inconclusive results, and can compromise the entire audit process. A comprehensive PSM systems audit at a large continuous process plant can require four to eight person-weeks of effort, including preparation and reporting.

The nature of a company's operations will influence decisions on the scope of the audit program in several ways. In some companies, location-wide PSM systems apply to several process units. Where this is the case, it is both practical and efficient to review PSM systems for the entire location. In other instances, a large location may have separate plants with independent PSM systems. In such cases, time considerations may dictate performing separate audits of the individual plants at different times. Furthermore, there may be certain types of operations for which particular process safety management issues have limited applicability. For example, at a warehousing operation there is limited applicability for process integrity issues. However, a facility or unit that has higher than average frequency of process incidents may require more attention.

The use of other management control systems (e.g., self-inspection or internal reporting) may also influence decisions on the scope of the PSM systems auditing program. Where there are many effective process safety management control systems, it is comparatively less important for the audit program to be frequent and broad in scope. Where the process safety management audit is relied on as a principal mechanism for providing process safety management feedback to management, it is important that the coverage be broad and the frequency higher. In making this judgment, it is important to differentiate truly effective management control systems from those that lack substance.

There is no single "correct" approach to defining the scope of an audit program. Rather, decisions on audit scope should be made within the context of the overall process safety management program.

1.3 Audit Frequency

The frequency with which PSM systems audits are conducted is dependent on the objectives of the audit program and the nature of the operations involved. Thus, audit frequency (i.e., the maximum interval between PSM systems audits) should be defined as part of the design of the audit program.

Among the factors to consider in determining audit frequency are degree of risk, process safety management program maturity, results of prior audits, incident history, company policies and government regulations. Each of the factors discussed briefly below should be considered in establishing audit frequency.

1.3.1 Degree of Risk

Degree of risk is the most important factor in deciding on the frequency of an audit. Generally the audit frequency will be higher for operations that pose higher levels of risk. Higher risks may result from the particularly hazardous nature of the materials present, the type of process involved (e.g., one that operates at elevated pressure), or the proximity of potentially exposed populations or resources.

1.3.2 Process Safety Management Program Maturity

The frequency of audits is likely to be higher for operations that have new or evolving process safety management programs, as compared with operations that have established, well developed programs. In the former type of operation, there is a greater chance for PSM systems to break down, either through the failure of individuals to implement process safety management programs consistently, or through the failure of the process safety management system to reflect the location's operating and management structure appropriately.

In a location with a more mature process safety management program, it is more likely that PSM systems have been integrated into the normal operations. As a result, less frequent reviews and verifications may be adequate. However, in such locations care must be taken to avoid complacency about the process safety management program.

Changes in either the process safety management program or the audit criteria may prompt reconsideration of established audit schedules. If a new program or new performance criterion is introduced, it may be desirable to perform an audit sooner than originally intended to verify program implementation.

1.3.3 Results of Prior Audits

When the results of an audit indicate significant gaps in process safety management system implementation, this may indicate the need to perform the next audit sooner than the program schedule would normally indicate. Companies that are developing process safety management programs sometimes conduct "baseline" audits to identify gaps in management systems and to focus their process safety management system development efforts.

1.3.4 Incident History

When a location has experienced frequent incidents or "near misses" (see Chapter 9), it may be appropriate to increase the frequency of the audits. In addition to identifying possible management system deficiencies, more frequent audits may increase awareness of process safety at the location.

1.3.5 Company Policies and Government Regulations

Company policies and government regulations sometimes specify a required audit schedule. For example, a corporate policy may specify that all manufacturing operations be audited every two years, while regulations on process safety management may specify that PSM systems audits be conducted at least every three years. In this case, a two year audit frequency would satisfy both requirements.

1.4 Audit Staffing

Conducting a comprehensive PSM systems audit normally requires a team effort. Involving a multiperson team in the audit process brings more than one perspective to bear, provides an opportunity for intrateam discussion of observations, and allows involvement of personnel with a variety of disciplines, skills, and experiences. While a limited-scope audit (e.g., addressing training for a process unit) can be conducted by an individual, most PSM systems auditing is performed by a team.

The ideal team for PSM systems auditing consists of no fewer than two and no more than six members. A single individual can conduct an audit, but the one-person approach lacks the benefit of bringing a variety of insights to the process. On the other hand, a team of more than six can be used, but teams of that size are more difficult to coordinate. The ideal team size for any particular audit depends on the size of the facility, the scope of the audit, and the amount of individual work (versus team activity) employed in the program.

The ideal team for a PSM systems audit will include individuals who have:

- familiarity with the process,
- experience in process safety management, and
- experience in audit techniques.

More than one of these characteristics may be found in a single team member.

There are a number of strategies that can be successfully used in building a PSM systems audit team. Each of these strategies offers its own advantages and disadvantages. Some are discussed below.

Staffing an audit team exclusively with individuals from the process unit being audited is generally not desirable. Using such staff would provide a team with a great deal of familiarity with site operations and personnel. However, this approach sacrifices the benefits often derived from having "fresh eyes" looking at a unit. In addition, the use of staff from the unit being audited can make it difficult to avoid potential "conflicts of interest," that is, instances where an auditor is reviewing things for which he has at least some responsibility or involvement, or where the auditor reports to the manager whose activities are being audited.

A variation on the above approach that offers some of the benefits, while avoiding some of the problems, is to use interfacility exchanges to staff audits. With this approach, an audit team would be comprised of individuals from other locations within the company where similar operations are performed. This would provide a team with a high degree of process familiarity, but with none having direct involvement in the operations of the plant being audited. This approach can also help facilitate technology transfer among locations.

However, the difficulty of freeing facility staff from their regular jobs to conduct audits at other facilities often means that an individual will only be able to participate in one or two audits per year. As a result, the audit team may lack members with strong audit skills.

Some companies employ a staff of dedicated auditors. Sometimes these staff members comprise the audit team, and other times they are used as team leaders with groups of facility staff made available through interlocation exchange. Dedicated auditors are best able to develop strong auditing skills, and develop a broad perspective on the topics being audited, because they see a wide variety of operations. However, they may not have in-depth process knowledge for all processes. In some companies, the dedicated auditor position is a pass-through job which is part of the staff development process. The use of a dedicated audit staff can help provide continuity when follow-up audits are done, but may lack the fresh perspective of someone new to a facility. A team that includes a mixture of dedicated auditors and temporarily assigned auditors can help preclude this deficiency.

Sometimes outside consultants are used in staffing audits. They may be asked to conduct audits as independent audit teams, to lead teams comprised of company staff, or to supplement the available internal staff working under the direction of an internal team leader. Outside consultants may provide a degree of independence to the audit process, and they may help supplement scarce internal resources. However, during an audit there is an opportunity to gain valuable knowledge about and appreciation for PSM systems. If outside consultants are used exclusively, the company may fail to capitalize fully on, and to enhance further, the knowledge of the internal staff.

Whichever staffing strategy is chosen, it is important that the audit team be trained. Effective auditing requires knowledge of both process safety management and of audit skills and techniques (see Chapter 2). While many people have one or the other, a team's effectiveness will be limited unless each team member has both types of knowledge.

1.5 Audit Reporting

At the conclusion of the audit, it is important that the findings be documented in an audit report. The report should be issued in a timely fashion, to expedite initiation of corrective actions. In designing the reporting process and executing the actual preparation of reports, there is a series of issues to consider, each of which is discussed below.

1.5.1 Report Content

The audit report should document the results of the audit, indicating where and when the audit was done, who performed the audit, the audit scope, and the audit findings.

The specific content of the audit report can vary. Some companies prefer a report that is an "exception report," addressing areas of deficiency and remaining silent on all other matters. Other companies prefer a report offering comment on every subject area reviewed, explicitly indicating the absence of deficiencies where this is the case. Some audit reports offer comments on areas noted as particularly strong, other reports only identify problems, while others include recommendations. Some audit reports prioritize findings, others employ a scoring system for results, while others do neither and simply list the findings.

The content of the audit report should be decided on as part of the audit program design, and should be consistent with the objectives of the audit program. There is no single correct way to determine the content of an audit report. However, it is important that once the content has been decided on, all audits produce reports that are consistent. It can be confusing and misleading for both facility managers and senior executives when different audit teams within a company include different types of information in their respective audit reports.

1.5.2 Distribution of Reports

When a PSM systems audit report has been prepared, it must be distributed to appropriate parties for follow-up action. Failure to distribute the audit report to appropriate individuals may compromise the value of the audit.

Distribution of the audit report may be determined by corporate policy. Typically, the appropriate recipients of the audit report include the manager of the facility being audited, and at least one level of supervision above that manager. In some organizations, the distribution may be more extensive.

Report distribution can be a sensitive issue in that an audit report will typically document deficiencies, which is sometimes seen by attorneys to be a poor practice. As a result, some corporate attorneys prefer to have audit report distribution managed by the law department. However, it is increasingly recognized that the value of self-diagnosis and corrective action is great, and that this value cannot be achieved without appropriate dissemination of audit findings.

This same sensitivity about the documentation of audit findings has sometimes led to the suggestion that audit findings be reported orally rather than in writing. That approach is not recommended as the sole means of reporting. To have an effective system for the resolution of audit findings and for tracking and follow-up of these actions, written reports are necessary. However, it is common for the audit team to communicate their findings orally to facility management before leaving the site.

1.5.3 Language of Reports

In writing audit reports, it is important that great care be taken to use appropriate wording. An audit report must clearly communicate the findings and observations of

TABLE 1-2
Examples of Appropriate Report Phrasing

Do not say	When you mean
The plant does not have	We were unable to confirm that . . .
	We were unable to determine that . . .
	The audit team was not able to verify . . .
	Plant personnel were unable to locate copies of . . .
I found _____ to be true	We understand that . . .
	We were told that . . .
	It appears that . . .
The plant is in compliance	On the basis of our review, we observed the plant to be in compliance
	On the basis of our review, it appears that . . .
	On the basis of *X* records examined, it was found that . . .

Source: Arthur D. Little, Inc.

TABLE 1-3
Examples of Audit Reporting Language to Avoid

alarming	gross negligence
appalling	incompetent
careless	intentional
criminal	neglect
dangerous	perjured
deliberately	reckless
deplorable	serious problem
dishonest	terrible
disorderly	violation
fraudulent	willful misconduct

Source: Arthur D. Little, Inc.

the audit team. However, it should be worded carefully so as not to imply findings or observations that go beyond those intended.

In an audit report, facts should be reported clearly and concisely. Every statement should be supportable; speculation must be avoided.

Tables 1-2 and 1-3 provide guidance on language to avoid in audit reports, and examples of appropriate report phrasing.

1.5.4 Report Retention

It is desirable to have an established policy on the retention of audit reports and backup records (including working papers and follow-up correspondence). Some companies have adopted policies calling for permanent retention of all records, while others retain records for a limited time (such as for seven years, or until completion of the next audit).

Retaining audit reports for at least one full audit cycle is important, as comparison with prior audits is a useful step during subsequent audits. In addition, regulatory requirements may mandate record retention.

1.6 Audit Follow-up

Subsequent to the audit itself, a follow-up and corrective action stage must occur. An audit will generally identify areas in which improvement is needed. The timely implementation of corrective actions will enhance process safety.Following issuance of the audit report, an action plan should be developed. This plan should include the timetable for implementing follow-up actions, and the person responsible for each indicated action. Accordingly, the action plan represents both a project schedule for the follow-up activity, and a management control document which can be used to monitor the status of corrective action. Where the facility specifically decides that no action is necessary on an audit finding, this should be noted and the reasons explained, lest subsequent reviewers perceive that the finding was ignored.

The action plan should be developed by the manager(s) responsible for the facility or operation that has been audited. This individual is responsible for process safety management at the facility, and must take responsibility for enhancements based on audit results. There should be an established system for review and approval of the action plan by appropriate levels of management.

Copies of the action plan should be distributed to everyone assigned responsibility under the plan, to the audit team, and to the next higher level of management. In most programs, the copy sent to the audit team is for information and to aid in subsequent verification. The role of the audit team may or may not include approving or evaluating the action planned, depending on the individual company's organization and assignment of responsibilities.

On a regular basis, the action plan should be updated to indicate which items are complete and the status of other items. As items are completed, the specific action taken should be documented and kept on file. Quarterly updating of action plans is often used, but more or less frequent updates may be chosen. This audit follow-up ensures that the company is documenting its intent to address audit findings, and provides assurance to management that appropriate steps are being taken.

Responsibility should exist within the organization for tracking action plan status. In some organizations this is part of the audit program. In other organizations, this is seen as a line function and is performed by the facility management. It is important that tracking occurs so that slippage in implementation of corrective actions can be identified promptly. It is also important that, regardless of who actually conducts the tracking, line management assume its responsibility for the execution of corrective action plans.

A final step in the overall audit process is the verification of corrective actions. This is generally a role performed by the audit teams. It is important to have an independent verification that corrective actions have been undertaken and that these actions effectively address the audit findings. In some audit programs, verification of corrective actions is performed as part of the next regularly scheduled audit. In other programs, verification is performed sooner as a separate, special review. In either case, the same audit techniques used for verification of program implementation should be used periodically to verify the reported status of the action plan.

1.7 Quality Assurance

Quality assurance is an important issue in a PSM systems audit program. Those being audited and those relying on the results reported must have confidence that the program is being carried out in a consistent and thorough manner.

The development of performance criteria for the audit program is one method of helping to assure quality. Criteria for an acceptable audit often evolve as the audit program develops. The types of things included in performance criteria for a PSM systems audit program might cover parameters related to the team composition, the nature and number of facility staff interviewed, and the availability of key records to the audit team (past incidents, relief valve test records, etc.).

Independent review of the audit process is another quality mechanism sometimes used in audit programs. This may be done during or after the audit itself. In some programs, an independent quality assurance person accompanies the audit team on some fraction of the audits to observe the audit process. In other cases, the audit working papers and report are reviewed by someone who was not involved in the audit to provide a second check for accuracy and completeness. The independent check need not be performed by someone external to the company, merely by someone not involved in the audit being reviewed.

Periodic critiques and evaluations of the audit program can be helpful in identifying program weaknesses. Such reviews can be performed by a task force comprised of employees not involved in the audit program, by the company internal audit function, by a group of external peers (e.g., an auditor from another company), or by an outside consultant.

1.8 Summary

The design of a PSM systems auditing program requires a number of choices on issues such as scope, frequency, staffing, reporting, follow-up, and quality assurance. While there is no single best way to structure a program that will be uniformly effective for all organizations, it is important to clearly define program objectives and settle on a consistent approach before beginning the audit.

2

Audit Techniques

2.1 Overview

A number of basic activities are common among most audit programs. Some activities are undertaken before the on-site audit, some during the audit fieldwork, and others after the fieldwork has been completed. Virtually all audits involve gathering and analyzing information, evaluating PSM systems against criteria, and reporting the results to management. A team approach is commonly used to conduct these activities. *Plant Guidelines for Technical Management of Chemical Process Safety* (CCPS 1992) also provides information on audits of process safety management systems and includes examples of auditing questions and forms, which may be helpful.

Figure 2-1 presents the key steps in one approach to auditing. Most companies include each of these steps in their audit process.

2.1.1 Pre-Audit Activities

The audit process typically begins with a number of activities before the actual on-site audit takes place. These pre-audit activities include the selection of the facilities to be audited, the scheduling of the audits, the selection of the audit team, and the development of an audit plan, which includes defining the scope of the audit, selecting priority topics, modifying the audit guides, and allocating audit team resources. They may also include an advance visit to the facility to gather background information and/or administer questionnaires, or a request to the facility for background information. Although the exact timing may vary substantially from company to company, most established audit programs will provide for these steps.

SELECTING FACILITIES AND UNITS TO BE AUDITED

Determining appropriate audit scheduling and review frequency depends on the specific goals of the audit program and the number of facilities and functional areas included within the scope of the program. Facilities or functional areas can be selected by a number of methods; for example, random selection, potential hazards, or the importance of the facility in terms of business considerations (see Section 1.3).

SCHEDULING THE AUDIT

One of the major considerations in scheduling the audit is whether this is the first audit at the facility or by the corporation, since required lead time is always greater for the first audit. The scheduling process generally begins with the audit team leader communicating to the facility manager that the facility has been selected for a PSM

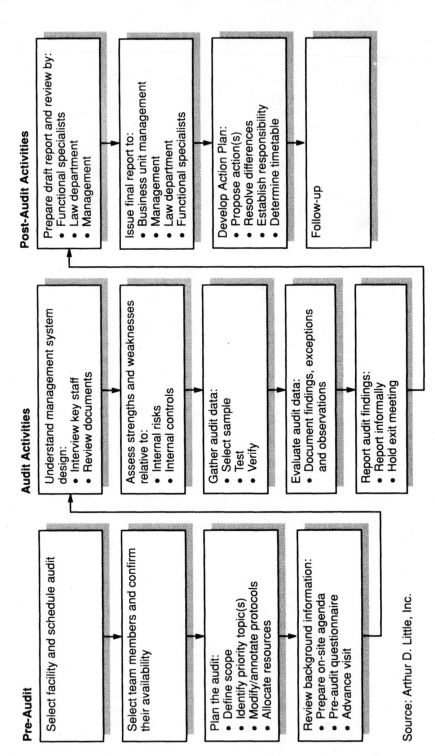

Pre-Audit

Select facility and schedule audit

Select team members and confirm their availability

Plan the audit:
- Define scope
- Identify priority topic(s)
- Modify/annotate protocols
- Allocate resources

Review background information:
- Prepare on-site agenda
- Pre-audit questionnaire
- Advance visit

Audit Activities

Understand management system design:
- Interview key staff
- Review documents

Assess strengths and weaknesses relative to:
- Internal risks
- Internal controls

Gather audit data:
- Select sample
- Test
- Verify

Evaluate audit data:
- Document findings, exceptions and observations

Report audit findings:
- Report informally
- Hold exit meeting

Post-Audit Activities

Prepare draft report and review by:
- Functional specialists
- Law department
- Management

Issue final report to:
- Business unit management
- Management
- Law department
- Functional specialists

Develop Action Plan:
- Propose action(s)
- Resolve differences
- Establish responsibility
- Determine timetable

Follow-up

Source: Arthur D. Little, Inc.

FIGURE 2–1. **Typical Steps in the Process Safety Management Audit Process**

14

systems audit. In selecting dates, the auditor's major considerations are to make sure the key facility personnel are available during the audit and to pick a time period during which the facility is in a normal mode of operation. About the same time the team leader schedules the facility visit, she should confirm the availability of the selected team members, and have replacements available in case of scheduling conflicts. The facility should be asked to designate a contact person to coordinate the collection of background material and the scheduling of interviews. The team leader should inform all audit team members about specific safety rules that the audit team will need to comply with, for example, safety equipment, facial hair, or clothing.

GATHERING AND REVIEWING BACKGROUND INFORMATION

Soon after the schedule has been established, the audit team leader should identify the types of information needed or desired in advance of the audit. Some of this information may need to come from the facility; other information may be available from other sources. Table 2-1 provides typical background information gathered in the pre-audit process. Only information that the audit team will review prior to the audit should be requested.

An audit questionnaire is sometimes administered in advance of the audit to assist the auditors and the facility in audit planning and preparation. This helps the audit team identify and understand key elements of the facility's internal process safety management procedures and systems, and identify audit topics that are not applicable at a particular facility. (More information is provided about the questionnaire in Section 2.2.2.)

TABLE 2-1
Typical Background Information Gathered in Pre-Audit Process

Previous Audit Report. A copy of the previous audit report
Action Plan. Status report on the resolution of previous audit exceptions
Regulatory Requirements. Copies of applicable federal, state, and local regulations
Corporate Policies. Copies of applicable corporate policies, standards, and guidelines
Facility Policy Manuals and Plans. Copy of tables of contents from current facility safety
     manuals, emergency plans, and other documents covering policies, procedures, and
     reporting requirements
Facility Organization. Current facility organization chart annotated to illustrate line and staff
     responsibility for all process safety areas under review, and to identify key contact people
Facility Layout. Maps or diagrams to illustrate location of different operations and of process
     safety and control system components
Process Descriptions: Discussion of process operations, flows, chemicals and control systems.
Selected Facility Reports. Copies of selected process safety status reports, self-audits, or
     other applicable reports
Incident Reports. Investigation reports for recent episodic incidents
Completed Questionnaires. Initial facility responses to specific questions asked during
     administration of questionnaires or other type of checklists

---

ADVANCE VISIT TO THE FACILITY

A facility visit prior to the audit can increase the effectiveness of the audit, particularly early in the development of an audit program. The objectives of a pre-audit facility visit are

- to inform the facility manager and staff about the audit program goals, objectives, and procedures, and
- to obtain information about the facility that allows the audit team to develop a more comprehensive audit plan.

The costs and benefits of an advance visit should be considered in the context of the goals and objectives of the audit program.

Whether or not there is a pre-audit visit, all information gathered should be reviewed by the team leader and team members, as appropriate, prior to arriving on site. The result of this pre-audit review is generally a list of questions and issues to be used in developing the audit plan. Experience suggests that a comprehensive review of background information will minimize the risk of omitting an important step as the audit plan is developed and then modified.

DEVELOPING THE AUDIT PLAN

An audit plan is an outline of what steps need to be done, how each is to be accomplished, who will do it, and in what sequence. Commonly, some form of audit "protocol" (see Section 2.2) serves as the outline for the audit plan. Prior to the audit, the team leader should identify the priority topics for review, modify and annotate the audit protocols or checklists as necessary, and make an initial allocation of audit team resources.

As part of the planning process, the audit team should determine an appropriate cross section of employees to interview and when they plan to conduct the interviews. This will help assure that the appropriate staff are available to the audit team. In addition, if there are specific things that the team will want to observe (e.g., an intermittent operation or an off-shift inspection of facilities), it will be helpful to make arrangements in advance.

2.1.2 Audit Activities

The first activity that normally occurs when the audit team arrives on site is an opening meeting. This provides an opportunity for the audit team and the key facility or unit staff to meet, and allows the audit team to explain their objectives, approach, and the overall audit process. The facility staff can then provide an overview of site operations. The opening meeting also provides an opportunity for the audit team to be informed about site safety rules and procedures, and for any special sensitivities to be aired. After the opening meeting, the audit team typically is given a facility tour.

The on-site audit typically involves five basic steps:

- Understand management systems
- Assess strengths and weaknesses
- Gather audit data
- Evaluate audit data
- Report audit findings

UNDERSTAND MANAGEMENT SYSTEMS

Most on-site activities begin by developing a working understanding of the facility's internal PSM systems. Sometimes auditors are tempted to rush into verification by reviewing documents; however, an audit should start with a thorough understanding of the management systems. The process safety management system involves the set of formal and informal procedures and activities used by the facility to control and direct process safety. Where informal management systems are in place (e.g., systems with little or no written documentation or procedures), the auditor must assess whether this form of management can be effective if operations or facility staff change.

This step usually includes developing an understanding of facility processes, internal controls (both management and engineering), plant organization and staff responsibilities, compliance parameters and other applicable requirements, and any current or past problems. Building on the information reviewed in advance of the site visit, the auditor's understanding is usually gathered from multiple sources, including interviews and review of documents. This understanding of management systems is usually obtained from management interviews, and later confirmed in discussions with other staff. The auditor records his understanding in a flow chart, narrative description, or some combination of the two in order to have a written description against which to audit.

ASSESS STRENGTHS AND WEAKNESSES

After clearly understanding how process safety is intended to be managed, auditors then evaluate the soundness of the facility's management systems to determine whether they will achieve the desired performance when functioning as intended. For each of the areas or topics assigned, the auditor should ask herself, "If the facility is doing everything the way they say, is that acceptable—will the facility be in compliance with applicable requirements, and is the company adequately protected?"

In assessing the strengths and weaknesses of management systems, auditors typically look for such indicators as written policies, programs, procedures and plans, clearly defined responsibilities, an adequate system of authorizations, capable personnel throughout the organization, administrative controls, documentation of actions, and internal verification. It is far easier to identify significant weaknesses in internal controls than to determine adequacy. Each of these indicators usually requires significant judgment on the part of the auditor since there are no widely accepted standards an auditor can use as a guide to what is acceptable internal control. Thus, many auditors look to the audit program objectives, as well as to the corporation's basic process safety philosophy, for guidance about what is satisfactory internal control.

This step helps determine how the balance of the audit will be conducted. Where internal controls are judged to be sound, the auditor will spend time confirming the existence of the control systems and testing whether they function effectively on a consistent basis. On the other hand, if the design of the internal system is judged inadequate to ensure the desired results, this will be an audit exception.

GATHER AUDIT DATA

Audit data are collected by the audit team to verify and validate the functioning of PSM systems. This forms the basis on which the team determines whether the process safety management system has been implemented as designed. Data are gathered in many ways, including review of records and interviews with personnel at a variety of levels. Suspected weaknesses in the management system are assessed in this step. Also, management systems that appear sound are tested to verify that they work as planned and are consistently effective.

The means by which auditors collect data fall into three broad categories:

- Interviewing
- Observation
- Verification

EVALUATE AUDIT DATA

Once data gathering is complete, the data are evaluated to identify audit findings. A finding is a conclusion, positive or negative, reached by the audit team based on data collected and analyzed during the audit. Findings are reviewed in terms of process safety management system criteria to determine their significance. Negative findings are typically called exceptions.

Audit teams usually make preliminary evaluations of their data throughout the audit and compare notes at the end of each day. Most audit teams then devote a few hours at the end of the audit to jointly discuss, evaluate, and finalize these tentative audit findings. The audit team confirms that there is sufficient data to support all findings, identifies trends in findings that may be more significant than the individual deficiencies, and summarizes each finding in a way that most clearly conveys its significance. The auditor should be careful in reaching conclusions based on single data points, and should strive to confirm preliminary findings using other data sources. All findings should reflect the consensus of the team.

REPORT AUDIT FINDINGS

During the audit, process safety management system findings are discussed on a continuing basis as the auditors and facility personnel interact to avoid having any surprises at the final closeout meeting. Findings may also be summarized in daily "wrap-up" meetings. The formal reporting process usually begins with an exit or closeout meeting between the audit team and facility personnel. During the exit meeting, the audit team communicates all findings noted during the audit. Any ambiguities about the findings are then clarified and their ultimate disposition (e.g., for audit report, for local attention only, etc.) discussed. Some companies have their audit teams make recommendations in addition to, or instead of, findings.

2.1.3 Post-Audit Activities

After the on-site audit work is complete, the audit team must complete its report and sometimes monitor the completion of an action plan to address audit findings.

When the team has left the site, all audit findings should have been identified and communicated to the local staff at the closing meeting. The audit team usually prepares a draft report, has the report reviewed, and issues the final report.

Some companies prepare a draft of the audit report on site. Most, however, prepare a draft audit report shortly after the on-site audit is completed. This draft usually undergoes review and comment before a final report is issued. Each company will have its own review process for audit reports. In virtually all cases, the location audited has an opportunity to review the report at the draft stage. In many companies, reviewers include a predefined group which may include other experienced auditors (peer reviewers), functional specialists and attorneys. The purpose of the review is to assure that the report is clear, concise, and accurate, rather than to modify the audit team's findings.

Subsequent to issuance of the audit report, the audited facility or unit should prepare an action plan for resolution of audit exceptions. The action plan should indicate what is to be done, who is responsible for doing it, and when it is to be completed. The action plan is an important step in closing the loop, both ensuring and demonstrating that audit findings are being addressed. All exceptions should be addressed and any action taken to address the exceptions or the rationale for not taking any action documented.

The role of the auditors with respect to the action plan differs among companies. In some companies the auditors receive copies of the action plan as well as periodic (e.g., quarterly) progress updates and are responsible for tracking the resolution of exceptions. In other companies, the auditors receive a copy of the action plans simply to complete their files, and then have no further role (until the next audit). Auditors are sometimes asked to review the action plan to ensure it addresses the "intent" of the exceptions. While either approach can be effective within the context of a well-designed program, it is always the responsibility of operating management, and not auditors, to write and implement the action plan.

2.2 Audit Guides

The PSM systems audit is most commonly supported by some important tools: protocol, checklist, internal controls questionnaire, and/or topical outline. While there is considerable latitude in current practice, most audit programs use some form of these devices.

2.2.1 Protocol

An audit protocol is a written step-by-step guide for accomplishing the audit, developed as part of the audit program design and used during each audit. It typically includes a listing of specific audit steps and procedures that are to be performed to gather data about facility programs and their implementation. The audit protocol provides guidance to the audit team in the collection of data.

A standard protocol can be one of two types: "discretionary" or "fixed." The discretionary type of protocol lists all audit procedures and verification tests that could be used for achievement of the audit goals and objectives. The auditor uses it much like a menu, selecting those procedures appropriate to the specific audit, and documenting those results.

The fixed protocol lists a series of procedures that must be carried out in every audit unless there is good reason to deviate. Fixed protocols are used where audit goals are served by some degree of standardization from audit to audit. Here the auditor must carefully document the reason why certain audit procedures were judged inappropriate or otherwise omitted.

Whether a discretionary or fixed protocol is selected, a number of steps should be taken in developing the protocol:

1. Decide the scope of the audit, including process safety management elements to be covered, and specific topics within each functional area, then list the selected audit topics.
2. For each topic selected in Step 1, identify and list the performance criteria against which the locations will be audited. These may include regulatory requirements (federal, state, local) and/or corporate and facility policies and procedures.
3. Determine and identify the depth of review for each topic selected above. (For example, should the auditor examine all safety training records or a selected sample of training records?)
4. Determine the type and level of audit techniques (e.g., interviewing, observation, verification) to use for each topic selected, paying particular attention to audit resources and time constraints.
5. Prepare draft audit protocol from Steps 1 through 4.
6. Have the draft audit protocol reviewed for accuracy and completeness.
7. Revise and complete the audit protocol.

The audit protocol serves as a record of the original plan for conducting the audit and as a record of the auditor's performance against that plan. The completed protocol provides a record of audit procedures that were performed and documentation of the rationale for any modifications of the plan. Figure 2-2 provides an example page from a PSM systems audit protocol.

2.2.2 Questionnaire

Some PSM systems audit programs supplement their audit protocol with a questionnaire designed to assist in identifying and reviewing internal management procedures and systems. This "internal controls questionnaire" allows a large amount of background information to be collected quickly and efficiently, and assists in identifying audit items not applicable to a particular facility's PSM systems. An internal controls questionnaire generally includes questions aimed at identifying and understanding key elements of the facility's management systems and procedures (e.g., maintenance, record keeping, internal reporting).

When the internal controls questionnaire is being prepared, the audit protocol may be used as a guide in determining the types of information to request. The auditor must have a clear understanding of what information is needed when preparing the questionnaire, and know the objective, scope, and focus of the audit. The questionnaire should include the items necessary for the auditor to gain an understanding of the

---

Introduction
Background Information

1. Review background information obtained from the facility to develop a general understanding of process safety hazards, areas of process safety concern, chemicals and processes used, etc. Typical background information includes:

   a. List of major hazardous chemicals used at the facility and applicable MSDS sheets for those chemicals;

   b. Description of the facility including site plan, flow diagrams, and general arrangement diagrams;

   c. Types of operations conducted;

   d. Organizational charts;

   e. Environmental, Health, and Safety Policy and any other applicable corporate policies;

   f. Selected facility reports (i.e., insurance carrier reports, incident reports, etc.);

   g. Copies of prior PSM audits or any other safety audits conducted at the facility;

   h. List of process hazards analysis conducted; and

   i. Number of contractors typically on-site.

2. Obtain and review applicable company and facility plans, policies, procedures, and standards.

3. Obtain and review Pre-Audit Questionnaire completed by facility.

FIGURE 2.2. **Example page from an audit protocol Ffrmat (page 1 of 17).** (Source: Arthur D. Little, Inc.)

operations and systems within the facility. During the audit, the questionnaire can be used in determining specific items for follow-up. The information gathered in the questionnaire should be documented.

There are a variety of ways to administer the internal controls questionnaire. Most audit teams meet with facility personnel at the beginning of the audit, with the team leader administering the questionnaire to appropriate facility personnel. It may also be administered by the audit team leader to facility personnel in advance of the audit. A few companies send their internal controls questionnaire to the facility for completion and return to the audit team leader prior to the audit. This approach saves the cost of direct administration, but sacrifices the additional information that usually accompanies face-to-face communications and increases the chance for misunderstandings, (for example, due to differences in terminology).

2.2.3 Topical Outline

A topical outline is a short summary or list of the major topics to be covered during the audit. Figure 2-3 provides an example page from a process safety management topical outline. As a simple list of key subjects, the topical outline relies to a great extent on the experience and judgment of the auditor.

The principal advantage of a topical outline is that it is short and easy to use for someone who knows how to go about reviewing each topic. The principal disadvantage is that it does not provide any substantive guidance to the audit team. Because it does not specify the exact procedures or manner in which each topic is to be reviewed, it does not facilitate a review or critique of the auditing effort to ensure its quality, and it does not assure consistency among auditors or audit teams.

2.3 Gathering Data

During the on-site audit, each audit team member gathers data to evaluate the facility's PSM systems. Data gathering techniques are discussed in this section.

2.3.1 Data-Gathering Methods and Sources

While a more extensive classification of data gathering methods is possible (for example, observing, examining, questioning, analyzing, verifying, testing, investigating), the three basic approaches are interviewing, observation, and verification.

Interviewing is perhaps the most frequently used means of collecting audit data. Here, the auditor asks facility personnel questions, both formally (e.g., via a questionnaire) and informally (e.g., through discussions). Interviews and discussions often provide satisfactory explanations of unclear items in the records. In evaluating information gained through interviews and discussions an auditor should consider the following factors:

---

1. Process Safety Information

1.1 Written process safety information

1.1.1 Chemical hazard information
    1.1.1.1 Toxicity information
    1.1.1.2 Permissible exposure limits
    1.1.1.3 Physical, reactivity, and corrosivity data
    1.1.1.4 Thermal and chemical stability data
    1.1.1.5 Hazardous effects of inadvertent mixing
1.1.2 Technology of the process
    1.1.2.1 Block flow diagram or simplified process flow diagram
    1.1.2.2 Process chemistry
    1.1.2.3 Maximum intended inventory
    1.1.2.4 Safe upper and lower limits for temperatures, pressures, flows, and/or compositions
    1.1.2.5 Evaluation of the consequences of deviations, including those affecting safety and health
1.1.3 Equipment in the process
    1.1.3.1 Materials of construction
    1.1.3.2 Piping and instrument diagrams (P&IDs)
    1.1.3.3 Electrical classification
    1.1.3.4 Relief system design and design basis
    1.1.3.5 Ventilation system design
    1.1.3.6 Design codes and standards employed
    1.1.3.7 Material and energy balances
    1.1.3.8 Safety systems (e.g., interlocks, detection and suppression systems)
1.1.4 Documentation that equipment complies with recognized and generally accepted good engineering practices
1.1.5 Determination and documentation that equipment is designed, maintained, inspected, tested and operating in a safe manner for existing equipment that was designed and constructed in accordance with codes or standards that are no longer is use.

FIGURE 2-3 **Example Page from a Process Safety Management Topical Outline (page 1 of 12).** (Source: Arthur D. Little, Inc.)

- The level of knowledge or skill of the individual questioned concerning the topic
- The objectivity of the questioned party
- The consistency of each response with other audit data
- The logic and reasonableness of the response

As an auditor gains familiarity with facility operations and organization, he becomes more adept at choosing the right person to question and evaluating the answer. The auditor also can determine whether the answer is consistent with other responses or information received. Although the respondent may not intentionally deceive the auditor, it is human nature for facility managers and staff to want to describe facility practices in their best possible light.

In addition, facility personnel may have inherent blind spots or biases of which they are simply unaware. The reliance placed on data obtained through interviews will vary, based on the factors discussed above, but heavier weight generally is accorded information generated by other means. An auditor should seek additional information whenever she judges a person's response to be uninformed, biased, or otherwise unreliable. In crucial matters, the auditor should not rely on a single source of data but should obtain additional information from independent sources.

Observation, or physical examination, is often one of the most reliable sources of audit data. Where knowledge of specific operations or equipment is important, it is desirable for the auditor to observe them. In some cases it may be practical and desirable for the auditor to physically inspect the entire site.

Verification refers to the wide variety of activities that can be employed to increase confidence in the audit data and the facility's internal controls. Verification can be a very powerful technique in assisting the auditor to achieve the objectives of the audit.

Auditors often gather information through sampling a portion of a whole collection (population) of items. The methods by which they select the sample can affect the validity of the sample and of the conclusions reached. It is important to minimize sampling bias and to obtain as representative a sample as possible. The auditor should maintain control of the sample selection. More detailed information about sampling strategies and techniques is provided in section 2.3.3 of this chapter.

Audit programs vary regarding the amount and balance of interview, observation, and verification. Some programs depend on interviews as the primary means of gathering audit data. Many of the more sophisticated audit programs employ a considerable amount of verification to determine whether management systems perform as intended.

For each item to be audited, interviews typically take a matter of minutes, observation can take tens of minutes, and verification a matter of hours. Thus, the more items to be verified, the larger the resource commitment required. Almost always, more items could be verified (and more ways to verify each item are possible) than available audit resources allow. However, audits usually serve as a check on the PSM systems rather than as a substitute for it. Therefore, most audits do not examine every situation, item, or document.

2.3.2 Interviewing Techniques

The term "interview" is used to encompass the full range of oral communication throughout the audit process. In fact, any large volume of information compiled during typical audit "interviews" is usually gained through conversations with facility personnel.

Regardless of the setting, duration, and degree of formality, all audit interviews follow a common pattern: planning, opening, conducting, closing, and documenting.

It is important to keep in mind that interviewing is a dynamic rather than a mechanical process—one that will be somewhat different for each paired interviewer and interviewee. The following basic process should be helpful in establishing a framework for the overall process and increasing the effectiveness of the interviewer's on-site activities. The emphasis is placed on the interaction that develops between interviewer and interviewee rather than strictly on the activities of the interviewer.

PLANNING THE INTERVIEW

Prior to conducting the interview, the auditor should identify personnel to be interviewed, and outline what is to be accomplished and how she intends to maximize the effectiveness of the interview. During this time, several key points need to be kept in mind:

- *Iron out logistics.* Obtain a brief understanding of the individual's current title, responsibilities, and reporting relationships. Whenever possible, establish a specific time, place, and duration for the interview, keeping in mind the individual's other commitments and work schedule. Carefully consider where the interview will be conducted. Avoid settings that may be intimidating to the interviewee, and avoid the appearance of "ganging up," where several auditors are simultaneously questioning one interviewee.

    Whenever meetings or discussions are held with facility personnel, the auditor should take time to make the environment comfortable for everyone. If in an office or a conference room, he should rearrange the chairs if necessary so that all persons can sit in a comfortable position. It may be appropriate to close the door to eliminate outside noise and interference or to create an atmosphere of privacy

- *Define the desired outcome.* Before meeting the person to be interviewed, take time to identify the specific type of information desired or areas to be addressed.

- *Organize your thoughts.* Interviews need not be long to be effective. To maximize the effectiveness of the interview, organize and consolidate the questions to be asked. Many auditors find it helpful to jot down a list of questions prior to the interview and, if possible, quickly review the list with other team members to see if they can think of additional questions or a more logical flow.

OPENING THE DISCUSSION

Perhaps the most crucial aspect of any interview is the opening communication, both verbal and nonverbal. While the total duration of the "stage setting" may be brief, the quality of information gathered during an interview is closely related to the interviewee's sense of comfort. To build the desired sense of comfort and confidence, auditors should follow a few basic guidelines:

- *Introduce yourself.* The auditor should begin by introducing herself, explaining why she is at the facility, and briefly recap the purpose and scope of the audit.

She should also explain her desired outcome for the specific interview, and indicate what she hopes to learn.

- *Ensure appropriateness of time.* To enhance rapport, confirm with the interviewee that the time is convenient for him ("is this a good time?"), in order to minimize the chances of being cut short or interrupted. As part of this approach, inform the interviewee of the estimated amount of time likely to be needed.
- *Explain how the information will be used.* Explain that the primary purpose of the discussion is to help the auditor develop a complete understanding of how the facility manages its process safety activities, not to try to "test" or find fault with the interviewee's operating practices. Explain that specific individuals' comments will be kept confidential when findings are reported.

CONDUCTING THE INTERVIEW

After establishing a comfortable setting and some degree of rapport with the interviewee, the auditor should shift the emphasis to obtaining specific information. For example:

- *Request a brief overview of the interviewee's job.* Experience has shown that even if an auditor is seeking answers to two or three very specific questions, it is always desirable to begin each interview by asking the person to spend a few minutes explaining how she fits into the overall organization at the facility and what her principal responsibilities are.
- *Gather detailed information.* Probe for answers to specific questions, following the applicable steps in the audit protocol. To ensure that the information gained is useful, pay attention to the techniques discussed: concreteness, respect, and constructive probing.
- *Concreteness or specificity of response.* The most effective way to obtain specific and concrete responses is for the auditor to ask specific and concrete questions. Vague queries generally result in nonspecific responses that are seldom useful. The auditor must control the interviewing process both to elicit concrete answers and to limit the discussion to relevant issues. However, leading questions that suggest a desired answer should be avoided.
- *Respect.* There are few more direct communications of respect than the commitment of the auditor to understand the interviewee's responses. That is, the auditor should focus on the *information* being given while deferring critical judgments about the respondent. Inadequate or incomplete answers often do not indicate that the interviewee lacks the ability to respond adequately, but rather that she is anxious about the interview, or that the question is open to more than one interpretation. Helping the interviewee to clarify and/or deepen her responses communicates respect and interest and provides a vehicle for eliciting specific responses. Confirming the responses by paraphrasing them back to the interviewee shows interest in the information being offered, while also allowing the auditor to assure that he has understood properly.

- *Constructive probing.* Constructive probing is often necessary, especially when interviewees provide responses that are inconsistent or conflicting. When questioned about the apparent inconsistencies, respondents are usually able to explain them satisfactorily. It is important, though, that the auditor phrase inquiries to focus on the data rather than confronting the respondent; that is, the effect of the inquiry should not be to criticize the respondent for being inconsistent, but rather to enlist the help of the respondent in clarifying the information.
- *Summarize the information learned.* Prior to closing the interview, review the information learned with the interviewee to ensure that the data gathered are correct. On longer interviews, this should be done several times; likewise, when several topics are discussed, be sure to summarize the information gathered at the end of each topic. In summarizing, pay particular attention to distinctions or refinements that the interviewee offers in response to the auditor's summary.
- *Provide feedback, as appropriate.* Feedback may be requested by the interviewee at various stages in the interview process. Because policies may vary from company to company regarding making recommendations and suggestions directly to facility personnel, auditors should understand those policies prior to providing feedback to facility personnel.
- *Do not exceed the agreed-upon time limit without checking.* A statement such as "This is taking a bit longer than I told you it would" or "Would another ten minutes be okay?" would suffice.

CLOSING THE INTERVIEW

It is particularly important to close each interview in a concise, timely, and positive manner. To ensure that the interview is productive and effective, end on a positive note. To ensure this, thank the interviewee for his time (and cooperation, candor or insights, where appropriate). In this way, the auditor will not only set a positive tone for subsequent interviews if deemed necessary, but also help create a good impression of the entire audit team.

In concluding the interview or discussion, it is often useful to ask a question such as: "Is there anything else I should have asked you and haven't?" The auditor might also ask if the facility representative feels that the information given offers a fair and accurate description of the facility's efforts.

DOCUMENTING INTERVIEW RESULTS

The process of documenting interview results begins early in the interview, perhaps with a casual comment that the auditor hopes the interviewee does not mind if some notes are taken to help the auditor remember the information discussed. Then, immediately following the interview, take time to review working papers to ensure that they accurately and completely reflect the information obtained during the interview. Pay careful attention to the following:

- *Establish a context.* Record the name of the interviewee, their title or job responsibilities, the date and time of the interview, and the applicable protocol step that the interview addresses.
- *Take notes during the interview.* In general, note-taking in and of itself does not interfere with the interview process, and some experts even argue that many people are offended if notes are not taken. However, while note-taking may not interfere with the discussion, the auditor's behavior associated with the note-taking process may. The auditor should take notes throughout the interview, and should try not indicate areas of concern through note-taking patterns. It is important that the auditor help pace the discussion without obstructing the information flow. If more than one audit team member is present for the interview, note-taking and questioning can be a shared responsibility and an uninterrupted flow of discussion may be easier to maintain.
- *Flag the key items.* During and immediately after an interview, make sure to flag (by using tick-marks or some other means) important statements or observations to assure they are not overlooked.
- *Summarize the outcome.* At the end of the interview, look at the notes taken and review the key points recorded with the interviewee, and note any changes or additions as a result of this review.

2.3.3 Sampling Strategies and Techniques

Because auditing basically constitutes a check on, or verification of, the implementation of PSM systems at a specific location, audit team members generally take a "sampling" approach to examine large "populations" of records or documents or groups of employees to make a determination regarding compliance. Despite the fact that sampling is a well-established aspect of auditing, selecting appropriate sampling methods and sample size can be difficult. Thus, the auditor must exercise considerable caution when selecting a sampling method to gather information. If, for example, the sampling method does not adequately represent the population under review, the information gathered can be misleading and cause the auditor to draw a biased, inaccurate, or unsubstantiated conclusion. To help ensure that each sample selected is appropriate and defensible, auditors typically follow six basic steps:

1. *Determine the Objective of the Protocol Step Being Conducted.* What particular aspect of a regulatory requirement or internal policy will be reviewed? The answer to this question, although at times obvious, helps the auditor to identify clearly the boundaries of the population under review.

2. *Identify the Population Under Review.* What is the population of records, employees, etc., to be reviewed? What segments of that population are relevant to the audit? For example, when verifying the existence of a preventive maintenance program, the first step is to identify all equipment that potentially should have been

covered. Frequently, the size of the population can be estimated based on your review of selected documents, the observations made during reviews of PSM systems in place, and interviews conducted with facility personnel.

Independent records should be used whenever possible to develop the sample. For example, in reviewing training records it is not wise to start with a sample developed from a stack of training records provided by the facility coordinator. The training records available to the facility coordinator will only reflect those who have been trained (or, more precisely, those with completed training records). To gather data about the extent of training and training records, it would be more desirable to start with personnel department or payroll records and develop a sample of employees who should have been trained. Then, the training records could be reviewed to help determine whether each employee in the sample had been trained.

The final task in this step it to identify potential bias in the sampling frame. For example, consider the following questions: Was the auditor in control of selecting the frame of interest? From what records was the population under review identified? Are other data missing that would influence the sampling frame selection?

3. Determine the Sampling Method to Be Employed. Samples selected by an auditor are judgmental, but may be aided by a systematic selection strategy (see Table 2-2). A sample developed largely on the basis of the auditor's judgment may be appropriate where the size or nature of the population makes a systematic sample difficult or unreasonable to obtain. A systematic sample is one selected through the use of a systematic process chosen to represent the population that is being reviewed. Numerous methods are available to select a sample for review, but no one method is correct for all situations.

4. Determine the Sample Size. The appropriate sample size can be determined either on the basis of the auditor's judgment or statistically, depending on the goals and objectives of the audit program. In most audit situations, it may be desirable as well as adequate to review only 10–20% of the population. For very large populations, however, developing a sample size that represents 10% of the population may be too cumbersome or too time-consuming. In such cases, the auditor may want to select a smaller sample, but she should be sure that the sample is large enough to allow reasonable conclusions to be drawn, or otherwise be aware of the limitations inherent in drawing conclusions from the sample selected.

5. Conduct Sampling. Again, careful attention must be paid to any potential for bias entering the sampling. Independent records should be used wherever possible to develop the sample, and records for sampling should be selected by the auditor, not by facility personnel.

6. Document the sample, strategy, and methodology employed. To assure management that a reasonable audit was conducted and to ensure quality control of the

TABLE 2-2
Examples of Systematic Sampling Methods

• **Random:** Select items purely by chance

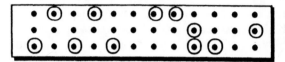

Select items in entire
population by chance (e.g.,
using a random number table)

• **Block:** Block sampling selects certain segments of the facility or employee population (e.g., months starting with J, records numbered 23-37).

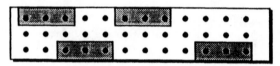

Arrange items by certain
segments or clusters and
randomly select items within
each cluster and compare to
other clusters

• **Stratification:** Involves arranging items by categories based on the auditor's judgment of risk, and then selecting a certain number of items from each category (new versus experienced employees, first shift versus third shift, etc.)

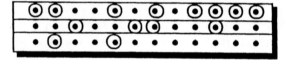

Arrange items by important
categories or subsets, then
sample within the groups

• **Interval:** Select every nth item

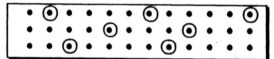

Select samples at intervals (every
nth item starting randomly).

---

sampling process, the auditor should be prepared to indicate why the particular sample was selected.

2.4 Evaluating Fieldwork

As fieldwork is completed, it is important to determine whether the information gathered by the auditor during fieldwork is sufficient to support the objectives of the audit and the conclusions of the auditor. Data in the audit process can be defined as whatever influences the auditor's findings and opinions. As such, the data obtained should be sufficient to force a conclusion to be drawn and enable that same conclusion to be drawn by different people.

2.4.1 Gathering Sufficient Information

Auditors frequently wonder whether they have collected enough information and the right kind of information to substantiate their understanding of a facility's key programs and management systems. Listed below are some tips for determining how much is enough. An auditor has probably gained enough information if:

- The auditor understands both management system design (e.g., facility policies and procedures) and implementation (e.g., availability of compliance records).
- The auditor has interviewed key personnel involved in conducting key functions or tasks, and can summarize to her satisfaction the basic programs, practices, and management systems.
- The auditor understands the probable cause(s) of any differences between management's and employee's perspectives, or between safety staff and operating personnel's perspectives.

2.4.2 Determining Adequacy of Information Gathered

The following four properties define the adequacy of information. The last requirement, persuasiveness, refers to its sufficiency. Additional information gathering may be necessary if these four properties are not satisfied.

RELEVANCE

Information gathered during a PSM systems audit should produce a flow of logic from the auditor's discoveries to the conclusions drawn. Thus, examinations of a sample of process change authorizations could constitute evidence that those process changes were handled appropriately in terms of the management of change requirements. However, this would not support the supposition that all changes within the facility have, in fact, been reviewed and documented.

FREEDOM FROM BIAS

Information used to reach conclusions must be free from any influence that would make one decision more attractive than another or that would exclude information supporting the alternative decision. Bias can arise from the source of the information or from the auditor's choice of items to examine. The answers received when interviewing management about their adherence to particular procedures may be biased, because it would be in management's best interest to appear competent and efficient. If an auditor decided to examine a random sample of available safety records without first determining that available records represented all transactions, the sample might be biased. Also, observations collected during a brief walk-around are likely to omit data points that are less accessible or out of the way and could, therefore, not be representative.

OBJECTIVITY

Objective data should lead two auditors examining the same information to reach the same conclusion. If, based on available information, two auditors reach different conclusions about a facility's compliance with particular requirements, then the information lacks objectivity and, therefore, is unreliable or insufficient for a decision, or the auditors may be biased and resolution is necessary to reach a decision.

PERSUASIVENESS

Information is persuasive when it forces a conclusion to be drawn and when that same conclusion is reached by different people. The persuasiveness may come from the volume of data, from the type of data, and from the source of data.

2.5 Working Papers

"Working papers" is the term used for the auditor's field notes and documentation which usually do not become part of the permanent audit record. Working papers are used in preparation of the final report, and are usually then destroyed.

2.6 Summary

The use of sound audit techniques is crucial to an efficiently conducted, thoroughly performed audit. Both the audit process and the techniques that support it should be carefully designed to meet the needs of the individual company. In addition, auditors should understand the objectives of the audit and the techniques that best achieve audit objectives.

3

Accountability and Responsibility

3.1 Overview

For a process safety management program to be effective, it is critically important that responsibilities be clearly assigned and those given responsibility be held accountable. The issues of accountability and responsibility must be addressed in reviewing every element of process safety management.

Since accountability and responsibility are principles rather than activities, they can be difficult to audit. However, there are specific indicators that show that process safety management accountability and responsibility are addressed within an organization. These may include

- a policy statement
- management commitment
- requirements for procedures
- individual performance measurement

In addition, it is important to consider issues that involve continuity of performance (i.e., systems to ensure that accountability and responsibility are maintained during the course of normal changes that may occur in operations and organizations).

3.2 Indicators of Accountability and Responsibility

The auditor should recognize that each organization will approach the establishment of accountability, goals, and objectives in a way that fits its own unique "corporate culture." There is no single best approach to these issues that will be uniformly effective for all organizations. The auditor should be open to alternative approaches, and be prepared to assess them in terms of their effectiveness, not their conformance to a preconceived "right answer."

3.2.1 Policy Statement

The policy of an organization with respect to process safety management should be articulated in a written document. In some cases, an individual plant may have a more detailed policy to supplement the general corporate policy. The more detailed policy statement might address process safety management policy along with other safety, health, and environmental issues, or it might be a separate process safety management policy. In either case, the auditor should be able to identify a statement of policy, assess its content, and determine whether it is up to date, how broadly it has been distributed, and its endorsement by management.

CONTENT

A process safety management policy should express an organization's commitment to safe design and operation. It should indicate that safe operation is a broadly shared responsibility that requires participation from every employee. It should express the policy of the organization regarding the role of safety, and should indicate the organization's approach to process safety management.

DISTRIBUTION

The process safety management policy should be broadly disseminated within the organization. This may be done through posting of policy statements and/or inclusion of policy statements in policy manuals and employee orientation programs, or other appropriate mechanisms. The auditor should ensure that the process safety management policy has been disseminated in a manner that is likely to be effective, given the organization's overall approach to internal communications.

MANAGEMENT ENDORSEMENT

The policy statement should be signed by the organization's senior management, indicating their endorsement of the principles embodied in the policy statement.

Senior management commitment to the process safety management policy can be demonstrated through concern shown during the facility tours, references made in written communications, or in speeches or other public statements.

UPDATING

The process safety management policy should be kept current as organizational responsibilities change and as policy objectives evolve.

3.2.2 Management Commitment

The auditor should verify that management is involved in appropriate ways in the process safety management program.

PERFORMANCE CRITERIA

Management's role begins with the establishment of performance criteria for the process safety management program. Criteria can be set at both the corporate and the facility level, and should be set for both process safety performance and the programs designed to effect that performance. Thus, in addition to overall performance criteria (e.g., 20% reduction of accidents), annual program criteria should be stipulated as well (e.g., revision of tank car loading operating procedures, or updating 20% of P&IDs). Performance criteria should be written and communicated to everyone involved in achieving them.

MONITORING PERFORMANCE

Management should be actively involved in monitoring performance and evaluating progress toward achievement of the performance criteria. There are many mechanisms that can be used for performance monitoring. The choice among them should be based on what works best within the management system of the organization. For example, management may require periodic written progress reports, periodic presentations to management meetings, or periodic assessments by a safety committee or safety supervisor.

The auditor should not be looking for any one particular approach to performance monitoring; instead, he or she should be seeking to confirm that monitoring is being done and that feedback is being provided by management to those responsible for programs.

RESOLVING DISAGREEMENTS

Management should serve as the "arbiter" when disagreements related to the process safety management program arise. Such disagreements may arise over the need for a mitigation measure, over the approach to take to mitigation, over the priorities to give different issues, or over the trade-offs among process safety and other organizational objectives.

In most organizations, staff are encouraged to resolve differences without involving senior management. However, an established mechanism should be provided for the resolution of disagreements to ensure that issues do not remain unresolved.

PROVIDING ADEQUATE RESOURCES

The management of any organization is charged with allocating limited resources among the competing needs of the organization. There should be access to appropriate expertise, and sufficient resources should be provided to implement process safety management programs. Decisions on resource allocation should reflect the need to address process safety management issues. The auditor should evaluate whether adequate resources are being provided to meet the performance criteria.

3.2.3 Requirements for Procedures

Procedures help to assure the clear assignment of responsibility. The auditor should sample process safety management procedures to ensure that they are complete as written and that they have been implemented in a comprehensive and consistent manner. During the review, the auditor should consider a number of aspects of the procedures, as described in what follows.

CONTENT AND LEVEL OF DETAIL

A procedure should be clear, concise, and comprehensive in its content. Wherever special terms or jargon are used, they should be clearly defined. All procedures should be written at a level consistent with the educational background of the intended users. Some procedures may be useful for training. These procedures should not presume knowledge of the operation being described. Where appropriate, cross-referencing to other procedures or programs should be included.

ASSIGNMENT OF RESPONSIBILITY AND AUTHORITY

The auditor should verify that each procedure clearly states who is responsible for implementing the procedure and who is to follow it, and that employees to whom the procedure applies are familiar with both their roles and the content of the procedure. Where procedures require approvals as part of their implementation, the individuals who are authorized to give these approvals should be specified in the procedure.

REVISION AND CHANGE

A system for writing approving, and effectively communicating revisions of procedures should be documented. The system should assure distribution of all updates to everyone with copies of the procedure.

The auditor should carefully review the revision and change system, probing for ways in which obsolete versions of a procedure could remain in place, or circumstances that could cause an employee not to learn about a change. Subsequent examination of various individuals' copies of a procedure that has been modified will serve to verify that the system for implementing changes is effective.

VARIANCES

Occasions will arise when variance from a procedure will be needed. When this occurs, it is important that a formal mechanism for reviewing and approving these exceptions be available. This should be addressed under management of change.

3.2.4 Individual Performance Measurement

The evaluation of individual performance relative to process safety management responsibilities is a powerful way to provide momentum to the program through the performance appraisal process. The auditor should verify that process safety management performance is included in performance evaluations.

3.3 Organizational Changes

Changes to an organization's structure and responsibilities should be reflected in process safety management responsibilities and accountabilities. The area in which these changes should be reflected are discussed below.

3.3.1 Responsibilities

When organizational changes involve the reassignment of operational responsibilities, process safety management responsibilities should be reassigned as well. It is important to avoid having important process safety management responsibilities left unassigned subsequent to organizational changes.

3.3.2 Performance Measurement

Individual performance measures should be modified to reflect organizational changes as they occur. It is important that process safety management performance measures be retained as organizational responsibilities change. The auditor should confirm the existence of systems that facilitate the revision of an individual's performance measures when necessary between regular review periods.

3.3.3 Resources

Some organizational changes coincide with changes in the staffing levels of an organization or facility. When this occurs, it is important to ensure that all process safety management activities remain assigned after reorganization, and that process safety management programs will continue to function effectively.

3.3.4 Procedures

Since procedures specify roles and responsibilities, an organizational change can result in the documents that cover them becoming outdated. For the procedures to remain effective, extensive updating may be required in the wake of an organizational change. The auditor should verify that there is a mechanism for updating.

3.3.5 Culture

Organizational culture can be an important contributor to an effective process safety management program. An inherent understanding of what performance is expected and what values should be applied in day-to-day decision-making can have a major influence on the results of an organization's process safety management program.

When an organizational change occurs, a corresponding change in the culture can occur as well. Changes in culture are not always undesirable, and they do have implications far beyond process safety. However, from a process safety management perspective, the effect of a culture change must be recognized. When a culture change occurs, it may call for changes to the PSM systems as well. The auditor should verify that PSM systems continue to function adequately within the new culture.

In case of team organizations, the individual responsible for the team activities should be identified. If there is no one person responsible for the team activities then the methodology that the team uses to manage process safety needs to be described.

3.3.6 Acquisitions

When a facility or business is acquired, it must be integrated into the PSM systems of the acquiring company. The auditor should verify that there are mechanisms to effect prompt and thorough process safety management system implementation.

Prior to the acquisition, the buyer should assess both the process hazards and the process safety management programs in place at the acquisition candidate. (If this assessment is not done prior to the acquisition, it should be done immediately after the acquisition occurs.) If gaps appear in the PSM systems, they should be prioritized and a schedule developed to address them. Responsibilities for integration of the acquisition should be assigned and its progress monitored. The auditor should confirm that this process has occurred.

3.4 Summary

The effectiveness of a process safety management program depends on the assignment of responsibilities, and on holding individuals accountable for their performance. Every element of process safety management requires assignment of responsibility and accountability.

Some aspects of auditing accountability and responsibility can be done separately, but most should be incorporated into the auditing of the other process safety management elements.

4

Process Safety Knowledge

4.1 Overview

Process safety *information* is the data describing the process and its chemistry. Process safety *knowledge* includes process safety information plus understanding or interpretation of the information.

Knowledge of the process is the cornerstone for the design and operation of a safe process facility. Without adequate knowledge of the chemistry and design of the process and how it is intended to operate, it is difficult to adequately implement many of the other process safety management elements (CCPS 1989; CCPS 1992). In particular, process safety knowledge is critical to

- conduct capital project safety reviews (Chapter 5)
- manage changes to the process (Chapter 6)
- implement a process equipment integrity program (Chapter 7)
- conduct hazard analyses and risk assessments (Chapter 8)

In addition, several of the elements of process safety management contribute to the body of process knowledge. For example, incident investigation, hazard analysis, and risk-assessment reports become part of the process knowledge, as do incident investigation reports. This contributes to a cycle in which process safety information is used during hazard analysis, after which the hazard analysis results become part of the process safety knowledge.

4.2 Audits of Process Safety Knowledge

An audit of process safety knowledge should include examination of process safety information. The auditor should first verify that there is a system to collect and maintain all the necessary process safety information described in Section 4.3 of this chapter. The next steps in the audit process are common activities for all process safety information. These include verifying that a system exists to ensure that data are accurate, reliable, and up-to-date, and that process safety information is available to all personnel who need to have access to it. Personnel responsible for updating process safety information should be interviewed. Potential users of process safety information should also be interviewed to assess the availability of information. Finally, the auditor should have discussions with project and process engineering managers and staff who would generally be responsible for initially assembling the process design package to assess their understanding of the process safety information system.

4.2.1 Data Sources

The usefulness of process safety information is dependent on the accuracy and reliability of the information.

The auditor should confirm that there are mechanisms for capturing information throughout the various stages of process development, design, construction, operation, maintenance, and decommissioning. Individuals involved in each stage of the process "life cycle" should receive guidance on the types of information to be documented, and where and how the information is to be retained.

Systems also should be in place to ensure that process safety information is accurate when placed in data storage systems. Whether process safety information is retained in paper form within files, in computer data bases, or on computer aided design (CAD) systems, the auditor should assure that information being entered is verified prior to storage. There should also be systems in place to identify gaps in the process safety information and a means to initiate programs to fill the gaps.

Process safety information systems should reference or identify the sources of data. This allows resolution of questions that may arise later, and facilitates clarification when necessary.

4.2.2 Data Availability and Distribution

The process safety management system should address information dissemination. Whether the process safety information is maintained in a central location in paper format (e.g., engineering department equipment files), or in a distributed form (e.g., process design reports), or in a multiuser automated data base, there are distribution issues, such as

- Access/confidentiality
- Data integrity
- Dissemination of updates

Staff involved in process safety should be aware of the types of process safety information available and where it can be found. When confidentiality is an issue, there may be reasons to restrict access to certain information relevant to process safety. In such instances, personnel involved in process safety should be informed of the existence of the class of information involved, and should be aware of the systems for gaining access to the information when required.

The management system should include controls over the changing of process safety information, to assure that the integrity of these data is not compromised. Data should be changed only by authorized individuals, and only in conjunction with appropriate quality assurance processes.

For those types of process safety information that are disseminated within an organization, there should be systems for assuring that only the most recently updated versions are in use. Approaches range from having every copy numbered and assigned to an individual, and requiring the return of acknowledgment when updates are sent · out, to using regular staff meetings to disseminate information about changes.

4.2.3 Maintaining Information

The process safety information management system should ensure that information is kept current throughout process changes, equipment maintenance, and other normal activities. Maintaining information requires appropriate linkages between the process safety information management system and the PSM systems for four other elements: capital project review (Chapter 5), management of change (Chapter 6), process equipment integrity (Chapter 7), and process risk management (Chapter 8). These linkages are depicted in Figure 4-1. As changes to the facility or process are reviewed and implemented, these changes should be reflected in the process safety information. Specific responsibility should be assigned and resources allocated for ensuring that this occurs.

4.3 Types of Process Safety Information

The following section describes many of the types of process information that can be important to process safety. The auditor should assure that information important to process safety is present and is maintained.

4.3.1 Chemical Data

A list of typical chemical data for the materials utilized as raw materials, intermediates, catalysts, products and waste streams in a process is given in Table 4.1. Some of this is information typically available on material safety data sheets (MSDS); other items might be found in technical literature or process development technical reports.

In addition to the chemical data listed in Table 4.1, information on regulatory exposure/reporting levels is also important. Examples of such information are listed in Table 4.2.

4.3.2 Design Data

Additional information such as process chemistry, catalysts, reactive chemicals, and kinetic data is usually required to perform the process design and develop what

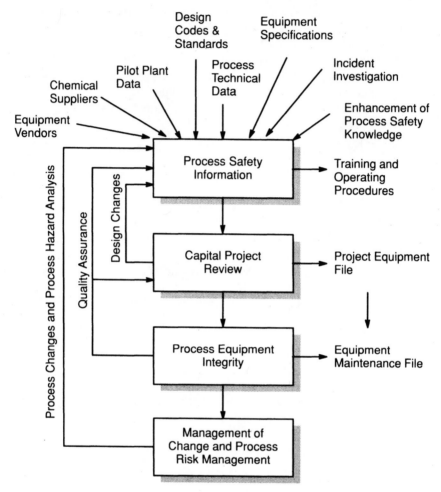

FIGURE 4–1. **Process Safety Knowledge Linkages**

is called the "front end" process design package. Such data are important for conducting a process hazards analysis. Examples of process design data are presented in Table 4.3. The type of process will determine which of these are pertinent.

4.3.3 Design Basis

The design basis is used during process design to define the operating conditions and limits of the process. Pressure and temperature ranges, flow rates to achieve the production capacity, concentrations and reactant ratios to provide safe and efficient

TABLE 4-1
Typical Chemical Data

Identification Data
 chemical name
 synonyms
 chemical formula
 code names
 trade names

Physical Property Data
 molecular weight
 density
 boiling point
 freezing point
 vapor pressure data
 viscosity
 refractive index
 surface tension
 solubility
 particle size and form
 azeotrope points
 critical pressure, temperature,
  volume
 vapor-liquid equilibrium

Thermodynamic Data
 latent heats of vaporization
 heat capacity
 thermal conductivity
 heats of fusion
 equilibrium constants

Reactivity/Stability Data
 flash points
 flammability data
 ignition energy for dust
 spontaneous decomposition conditions
 spontaneous polymerization conditions
 shock sensitivity
 pyrophoricity
 hydrophoricity
 reactivity with common materials of
  construction
 incompatibilities with other chemicals

Explosive Properties
 maximum rate of pressure rise
 peak explosive pressure

TABLE 4-2
Selected U.S. Regulations for Chemicals

OSHA Standards

EPA SARA Reporting Quantities

DOT Classifications and Placarding Requirements

EPA RCRA Classification

State or Local Requirements (e.g., California RMPP, New Jersey TCPA, Delaware EHSRMA)

TABLE 4-3
Typical Process Design Data

Process Chemistry
  equations for the primary and secondary chemical reactions
  equations for the important side reactions
  intermediate chemicals (especially if isolated during the process)
  sequence in which the reactions occur

Catalysts
  composition of the catalyst used
  properties of catalyst mixture
  competing side reactions
  hot spot potential
  catalyst inhibitors or quenchers

Reactive Chemicals
  compositions
  locations in the process
  contained in mixtures
  inhibitors and concentration required for stability
  runaway reaction data (e.g., accelerated rate calorimetry)

Kinetic Data
  main reaction rate constants
  side reaction rate constants
  decompositions
  auto-polymerization
  equilibrium constants
  heat of reaction

---

operation, and control philosophy are some examples. Also, the consequences of exceeding these limits should be defined.

Much of this information would be expected to be found in research reports or licensee data and is usually incorporated into the operating manuals. Although it is common practice to keep these reports in central libraries, copies containing these data should also be available to the site.

4.3.4 Process Flow Diagrams

Much of the process design information described above is shown diagrammatically on process flow diagrams (PFD). Process flow diagrams show major process steps (generally separate unit operations) such as

- Storage
- Reaction
- Purification

- Heating/cooling
- Compression/expansion
- Separation
- Mixing

Process flow diagrams can show information in varying levels of detail. The simplest of these is the block flow diagram. Each of the steps required for the process is represented by a separate block, and the major flows into and out of the block are shown. Principal temperatures and pressures for the steps may also be shown in the block. Block flow diagrams are usually used to represent conceptual designs. They are also used for showing the principal steps in a batch process.

A more detailed process flow diagram is generally used for representing continuous process units. The major pieces of process equipment are shown, along with the process lines into and out of the equipment, and principal utility demands. Operating conditions of flow rate, temperature, and pressure for the equipment or process lines are also shown.

For a cyclical process operation, or a batch process, the process flow diagram will be as described above except there are likely to be some supplemental sheets in which the major process steps are listed in the sequence they are performed. Along with this may be a valve position chart showing the valve positions during each step and the equipment that would be in operation.

4.3.5 Special Design Considerations

For some processes, there are special design considerations which must be met. Where applicable, these criteria should be well documented. Some examples of special design criteria are listed in Table 4.4.

TABLE 4-4
Examples of Special Design Considerations

Nonlubricated equipment
   —oxygen service
   —oil-free air
   —reactive gases
Exclusion of or requirements for oxygen
Exclusion of water
Flushing fluids for instrument systems
Intermediate nonaqueous heating or cooling media
Heat transfer fluids or salts
Exclusion of reactive contaminants (e.g. rust)
Shock sensitivity of materials
Reactivity with materials of construction (passivation)
Requirement for stress relieving of vessels or piping
Time/temperature sensitivity to thermal decomposition
Potential for brittle fracture
Lethal service requirements for vessels or piping

4.3.6 Piping and Instrumentation Drawings

Piping and instrumentation drawings show all process and utility piping, instrumentation and protective systems. These drawings also typically show design temperature and pressure ratings for vessels, piping specifications, relief valve setpoints, capacities of vessels and pumps, and heat duty of exchangers.

4.3.7 Plot Plans

A plot plan shows the layout of major process equipment on the site. This plan should note any special design and layout considerations, such as separation distances and underground pipeline locations.

4.3.8 Electrical Classification Plot Plan

This document is a plot plan of the unit or site showing the electrical classifications that are applicable to the different sections of the plant. It is important where flammable or combustible materials are handled.

4.3.9 Plot Plan of Underground Services

This drawing can be important to help avoid accidental rupture of underground services when excavating. Examples of services that are underground at some locations include:

- Process and raw material lines
- Water lines (process, cooling, potable)
- Fire water lines
- Process drain lines
- Electrical power cables
- Utility supplies (e.g., natural gas, nitrogen, air)
- Communication lines

4.3.10 Equipment Specification Sheets

Equipment specification sheets should be provided for each piece of process equipment such as pumps, compressors, tanks, vessels, heat exchangers, and on each protection system (e.g., overpressure protection, shutdowns, fire suppression). These sheets typically show the design codes and standards used, the expected process conditions, the design conditions, materials of construction, and other mechanical details.

Along with the equipment specification sheets, other information about the item should be available. Typical information might include hydrostatic test reports, manufacturing inspection reports, weld radiographs, internal inspections, and other documentation on tests and inspections conducted during fabrication and installation. These documents are key to ensuring equipment quality and mechanical integrity as discussed further in Chapter 7.

4.3.11 Piping Specifications

Piping specifications include, for each line, the type of pipe used along with data on the types of fittings, valves and gasketing materials that are acceptable to use for the expected process materials and conditions, and the applicable standards that the piping components must meet.

4.3.12 Safety-Critical Instrument Index

The safety-critical instrument index is a listing of all the instruments in the process which are important to maintaining process safety. These may include sensors, transmitters, controllers, control valves, pressure reducers, etc.

Each of the instruments should have a complete specification sheet with the design conditions, the required materials of construction for portions of the instrument in contact with process fluids, as well as the range it must cover and its protective function.

4.3.13 Electrical One-Line Diagrams

These drawings show the distribution of power from the source to all of the power users in the unit, including major transformer and switching apparatus.

4.3.14 Programmable Controllers and Computers

For a plant that is controlled or monitored by a computer, there are two sources of documentation that are critical. The first is the specifications for the hardware system, including the computers and the uninteruptable power supply (UPS). The second is the documentation of software—the coded instructions which control or monitor the operation of the plant when it is running.

Documentation of the computer system hardware usually consists of the service manuals provided by the computer system vendor. A second set of hardware data is the wiring diagrams showing the connection of field sensors into the computer system and those going out to the instruments which initiate process control action. Within the computer control system there are two elements that are subject to change and revision. The first is the configuration of the instrument control loops, control actions, tuning constants and alarm points. The second is the control software that will take process information, perform calculations, monitor for shutdown or unsafe conditions and take corrective actions. All of these functions of the control system must be under secure management control and not accessible to random changing by the operators.

Documentation of the alarm points, the configuration of the control loops, and the control software should be stored in a location separate from the process.

4.3.15 Vendor Data

Within a process plant there are many equipment items supplied as a package, complete with all mechanical and instrumentation components. Typical systems would be water softening or deionizing systems, furnaces, heaters or boilers, refrigeration units, HVAC systems, and air dryers. With these systems come sets of drawings, operating and maintenance instructions or manuals. These documents are essential to the operation and maintenance of the units and the securing of the correct spare parts for their repair.

The packaged units in the process should be identified in the process flow diagrams and the P&IDs. Since there are usually a limited number of copies of technical data for these units, at least one set of the vendor documents should be maintained in a central file with controlled access. Other copies may be found in the operations offices and the maintenance shops for reference when work or troubleshooting must be performed.

4.3.16 Other Information

The process safety information should also include reports documenting the process hazard identification and analysis activities that occur during the life of the process.

These include hazard analysis, audit, and incident investigation reports. These reports should be retained under an appropriate record retention policy.

4.4 Procedures

4.4.1 Operating Procedures

Most process safety information is used off line in design or modification of the process. Operating procedures represent a special type of process documentation in that they are required in day-to-day operations. Accordingly, the operating procedures should be in place for new and modified facilities prior to start-up.

Operating procedures should provide clear instructions for safely conducting activities involved in each process, and for switching between products in a batch operation. There should be procedures addressing each operating phase, including initial start-up, normal operation, emergency operations (including emergency shutdowns), normal shutdown, and start-up following an emergency shutdown or a turnaround. In addition, operating procedures should be developed whenever a temporary or experimental operation is to be conducted (see Chapter 6). For these and other routine procedures (e.g., loading/unloading, preparation for maintenance), a job task analysis may be done (see Chapter 10), and checklists may be developed.

In addition to describing the sequence of actions to be taken in operating the process, the procedures should describe process safety-related operating limits. For each of these limits, the procedure should make clear the consequences of a deviation, steps to be taken to avoid or correct a deviation, and safety systems in place to handle deviations outside acceptable limits.

The operating procedures should also describe safety information related to the process. For example, the hazardous properties of chemicals used or produced should be described. Where there are special inventory restrictions or quality control procedures associated with the process, these should be described.

The employees who operate the process should have ready access to the operating procedures and they should be familiar with the contents. The procedures should be written in language and terminology clearly understandable to process operators.

It is important that operating procedures be kept up-to-date. There should be systems in place for ensuring that applicable process changes are incorporated into the operating procedures (see Chapter 6), and that the procedures are periodically reviewed for accuracy and completeness.

The auditor should sample operating procedures to verify completeness. He should identify recent process changes or capital projects to ensure changes are reflected in the procedures. He should interview operators to determine their understanding of the procedures, particularly relative to training of new employees or start-up of new projects.

4.4.2 Other Procedures

The auditor should also be able to find procedures for other operations such as maintenance, emergency response, and training. There should be systems in place to assure the updating, dissemination, and accuracy of these procedures.

4.5 Enhancement of Process Safety Knowledge

Codes, standards and guidelines for design, fabrication, construction, operation and maintenance of process equipment change with time as better methods of analysis are developed and as criteria regarding acceptable risks change. It is important for a facility to keep abreast of advances in process safety knowledge.

The auditor should determine if there is a system to allow continuing education in process safety information for technical personnel. These personnel should read professional journals, attend conferences and maintain liaison with other professionals with responsibility for similar process facilities. The auditor should verify that this is taking place by exploring what the facility is doing relative to recent developments in process safety such as the AIChE Design Institute for Emergency Relief Systems

(DIERS) technology, new regulations at both the state and federal levels, and results of incident investigation reports from other facilities and companies.

4.6 Summary

Process safety knowledge is important in the design, fabrication, construction, operation, maintenance and decommissioning of processes. There should be systems in place to accumulate process safety information and communicate it to employees who need to use this information. As changes to the process occur, systems should be in place to ensure that process safety information is updated.

5

Project Safety Reviews

5.1 Overview

Changes of various types occur frequently in an operating facility. Changes which involve capital expenditure are capital projects. Most companies require formal approval of capital projects based on dollar value. As dollar value increases, the authorization level in the company also increases. Small projects may be approved by local management and implemented by facility staff. These small plant level projects may not fall under the company's formal capital project procedure. The auditor should be aware of the company criteria for defining a capital project, since all other modifications would need to be reviewed under the management of change procedures (see Chapter 6).

Since capital projects may involve equipment and/or technology changes, one or more formal safety reviews should be performed. However, the scope and complexity of capital projects range from installing a new pump to a major plant expansion involving new chemical reactions and equipment. The number and type of project safety reviews should be appropriate to the size, complexity, and nature of the project. Because project safety reviews can employ different techniques (e.g., hazard and operability study, failure mode and effects analysis, "what-if," etc.), auditing this process safety management element should involve someone with a reasonable understanding of project stages and the applicability and effectiveness of the various hazard evaluation methods.

The audit of this element should examine both the *presence* of an adequate program and the proper *execution* of the program. The former considers the existence and quality of the project safety review procedure. The latter addresses whether the procedure is being effectively implemented, given an understanding of how it is supposed to be implemented.

As indicated in Figure 5-1, project safety reviews interface with many other process safety management elements. For example, process safety knowledge and documentation are essential for a credible and effective review. Likewise, the accountability for major project decisions and risk management needs to be clearly understood by all participants. This requires strong coordination and communication among audit team members who may be auditing the other elements.

Depending on the findings of the project safety review, recommendations may involve a number of process safety management activities ranging from investigation

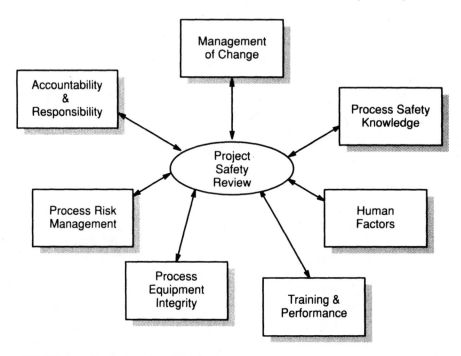

FIGURE 5–1. **Project Safety Review Interfaces**

of materials compatibility to labeling of control room alarms. Because of this, the auditor may need access to information from several departments when verifying follow-up and closure. Interviews with the safety coordinator, process chemist, engineering department supervisor, project engineer, technical service supervisor and engineer, operations manager, unit supervisor, or operators might be appropriate.

Another important area for the auditor to investigate is how project safety reviews are scheduled and funded. Project safety reviews should be incorporated into the project schedule and project control plan, and the cost and time of these reviews should be allocated. Adequate time and funding for any modifications resulting from these reviews should also be factored into the project schedule and capital appropriation. Lacking this, the project safety review system may be ineffective and largely ignored by the project manager. Therefore, the auditor should examine how project review costs and mitigation costs are estimated and allocated, and if the project safety review is integrated into the project schedule.

5.2 Project Safety Review Procedures

The auditor should begin with a review of the documentation that describes the project safety review procedure. This information may be located in the capital project procedures. There should be written procedure indicating

- When project safety reviews are to be done
- How they are initiated
- Who assigns staff for the review
- What techniques are used
- How the review is documented
- How responsibility for follow-up is assigned
- How follow-up is verified and documented

A copy of the project safety review procedure may be obtained during the pre-audit planning stage. Reviewing the procedure in advance has certain advantages:

- It provides the auditor with an understanding of the intended scope and approach to project safety reviews
- It establishes the date of the last revision of the procedure
- It provides insight about the management controls and responsibility for the system

An understanding of these aspects is needed to help focus subsequent audit activities.

Copies of the project safety review procedure should be readily available to affected company staff. The auditor should determine if all personnel who participate in project safety reviews have access to a copy.

Documentation should be addressed in the project safety review procedure. The auditor should review the reporting requirements, and review a sample of reports to determine compliance and quality of the project safety review. Some aspects to evaluate include

- Report type (interim draft, final)
- Content and format
- Distribution (recipients, copies, etc.)
- Archiving

The project safety review procedure should be reassessed periodically and revised as necessary. The auditor should establish whether there is a site/facility procedure for periodic updating and if it is effective. The frequency of the periodic procedure review should be determined. An interval of 1–3 years is typical.

There are two aspects of responsibility for keeping procedures current: that of initiating the periodic reassessment, and that of performing and issuing the update. The auditor should examine whether there are checks and balances in the management system to ensure that the procedure is reassessed in a timely manner. For example, having the person responsible for process safety management trigger periodic procedure reassessments and verifying completion is one way of providing oversight of the system. With regard to the actual review and re-writing of the procedure, the auditor should ascertain whether people who use and are affected by the procedure are involved in the reassessment/updating process. Interviews with the engineering and safety personnel should include a discussion of the updating process.

Because the project safety review procedure may affect several departments, there should be a formal sign-off requirement for changes to the procedure. The approvals might include plant manager, operations manager, engineering manager, and the individual responsible for process safety management. Some companies may also have the corporate safety function or other corporate staff involved, to ensure consistency among the various locations. In other companies these procedures may be developed and reviewed at the corporate level.

5.3 Hazard Analysis

Project safety reviews may occur at different stages of a project, and the scope of each review will depend on what phase the project is in. The complexity and size of a project will determine what safety reviews are needed, when they should be done, and how they should be done. Projects progress through several different phases, with different types and amounts of information available at these different phases. While not every project passes through every phase, one way to characterize these phases was presented in *Guidelines for Technical Management of Chemical Process Safety* (CCPS 1989):

- *Phase I*—Conceptual Engineering
- *Phase II*—Basic Engineering
- *Phase III*—Detail Design
- *Phase IV*—Equipment Procurement and Construction
- *Phase V*—Commissioning Prior to Startup

More information becomes available as the project progresses through the five phases. For example, process flow diagrams are typically available during Phase II—Basic Engineering, whereas during Phase III—Detail Design, the more detailed piping and
instrument diagrams will be available. Because more information becomes available, a more rigorous safety review can be done at Phase III than in Phase II. However, it is easier to correct problems when they are found during earlier phases. It is important to know which type of safety review to conduct at a specified phase in the project. Conducting an inappropriate type of review can waste time and money, while not providing adequate insight into the potential hazards. Therefore, along with assuring that a review is being conducted, an auditor should be able to verify that the type of review is appropriate.

A major part of the auditor's job is to assure that a good program on paper is actually being followed in the field. It is not sufficient to have a program in place if it is not being followed. Table 5.1 shows some of the criteria that should be used in evaluating the adequacy of the reviews conducted.

TABLE 5-1
Evaluation of the Common Elements of Project Safety Reviews

Element	Execution	Adequacy
Written Procedure	Procedure exists	Complies with accepted standards
Conductance	Review takes place	Follows accepted procedure
Documentation	Report completed	Follows accepted procedure and report is complete and usable
Response to findings	Changes to design or operation incorporated, or findings rejected	Provide independent proof that all findings were addressed or the rationale for rejection of findings is documented

5.3.1 Hazard Analysis Techniques

In addition to considering generic requirements of project safety reviews, it is important that the specific review technique being employed is completed in the proper manner. These techniques are described in *Guidelines for Hazard Evaluation Procedures: Second Edition with Worked Examples*. This reference provides a summary of generally accepted review methodologies, which will assist the auditor in assessing the adequacy of the technique being employed. Since the review techniques are sometimes loosely interpreted, the auditor needs to probe to find out how the hazard analysis is actually done. With the aid of the above-referenced book, the auditor should be able to determine which method is actually being used and if there are any steps missing.

The hazard analysis may reveal hazards that are not easily mitigated by engineered systems or procedural controls. Hazard analyses should consider the impact that a project may have on existing facilities. Management needs to understand the level of risk exposure before granting approval for continuing with the proposed design. In this situation, more quantitative techniques may be considered. The auditor should determine if there are provisions for quantitative risk assessment of major hazards and, if so, whether they have been used. A line of questioning to determine if risk assessment studies are ever done during project reviews should be pursued.

Safety reviews are often conducted during the conceptual engineering, detail design, and pre-startup phases. The conceptual engineering phase can involve both laboratory and pilot plant development work. The detail design phase may encompass basic engineering, detail design, and equipment procurement and construction. The pre-startup phase is the final stage before operation, which includes commissioning.

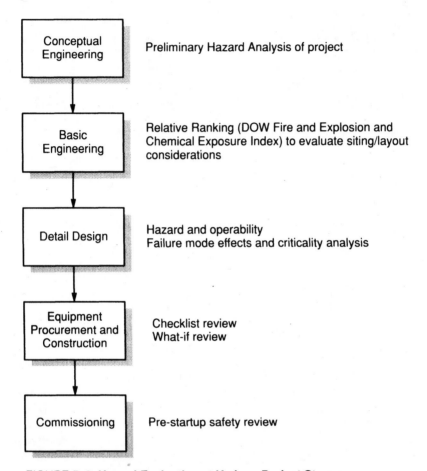

FIGURE 5–2. **Hazard Evaluation at Various Project Stages**

CONCEPTUAL ENGINEERING

The conceptual engineering phase can involve both laboratory and pilot plant development work to determine the process definition. The purpose of safety reviews at this stage is twofold. The first reason is to ensure that the laboratory and pilot plant systems are designed and operated safely. The second reason is to determine whether or not the project will be safe and environmentally sound when scaled to a full size operating plant. The auditor should understand the differences and verify that the safety review addresses both points.

At the conceptual engineering phase, Checklist, What-If, or Preliminary Hazards Analysis techniques are often used for hazard analysis. Checklists are generally used when dealing with known reactions and/or chemicals. These may be used to rule out testing of certain technologies or chemicals because of safety concerns. What-If and

Preliminary Hazard Analyses could be used for the same purpose when the processes under study involve new technologies and/or unfamiliar chemicals. It is not uncommon for a project to be canceled at this stage because of unresolvable safety concerns (e.g., lack of buffer zone between the plant and residential areas). An auditor should verify that appropriate reviews are required at the conceptual engineering phase before moving to the next step in the audit.

DETAIL DESIGN PHASE

In the detail design phase, all information from the conceptual engineering phase (including both safety and operability) is combined with basic engineering knowledge to design a process to satisfy the goals of the original project definition. The auditor should verify that safety information from the conceptual engineering phase was transferred and used. By talking to members of the design team and examining study reports, the auditor can determine if an adequate design review has been conducted. An essential part of this review is to verify that recommendations from the conceptual engineering phase review have been addressed.

After the piping and instrument drawings are available, a more in-depth hazard analysis can be completed (e.g., a Hazard and Operability study, Failure Modes and Effects Analysis (FMEA) or What-If/Checklist analysis). Regardless of what hazard analysis technique is used, the auditor should confirm that the safety review completed at this phase incorporated the piping and instrument drawings, the appropriate team members, and the correct follow-up.

PRE-STARTUP PHASE

Pre-startup safety reviews are very important because they are the final check before the process is put into operation. They include a review of the operating procedures and a walk-through inspection of the facility. Typically, final checks are made at this phase to verify that recommendations from previous reviews have been addressed, that all safety information (including that developed during the conceptual engineering and detail design phases) has been incorporated, that the constructed plant adheres to the intended design, and that operating, maintenance and emergency procedures are in place and operating personnel have been trained. The auditor should verify that the pre-startup safety review system meets these desired objectives and any plant-specific objectives.

Written protocols or procedures for project safety reviews should exist and should be reviewed by the auditor. The auditor should interview project safety review leaders, participants, and the ultimate customer (e.g., operations) to gain an understanding of how the review system is implemented. The auditor should also identify typical projects in progress or recently completed to audit the adequacy of the reports documenting the reviews.

5.3.2 Staffing

The project safety review must have an appropriate leader to organize and execute the review. There are three important criteria for selecting the chairperson: experience, leadership skills, and objectivity. The project safety review procedure may address minimum skill requirements for review leaders, such as educational background, years of appropriate experience, and project safety review training. With regard to the latter, there may be some trade-off between formalized training and participation as a team member. Another criterion that can be employed is type of process experience. For example, it would be preferable to have a leader with batch process experience to lead a process review of a batch reactor system.

The process safety review procedure should also address the objectivity of review leaders. There is a trade-off between knowledge of the process and the perception of objectivity that must be considered, and the project safety review procedure should consider this with respect to the selection of a review leader.

The project safety review procedure should also address the selection of review team members. The objective is to have a team composed of varied disciplines that encompass the skills and experience needed to conduct an adequate review, and enhance the team member interaction. Therefore, the project safety review procedure should indicate the preferred disciplines for the core team. These may be supplemented during the review by other members with specialized knowledge such as computer control, process safety, or maintenance. With regards to implementation, the auditor should verify that team members are selected on the basis of skills/experience, not purely job title.

The team includes experienced personnel, typically as follows:

- Team leader
- Team secretary
- Experts in fields relevant to the project (e.g., operations, maintenance, and safety)

5.3.3 Roles/Responsibilities

The auditor should verify that the project safety review procedure specifies the proper roles and responsibilities for team members. Responsibility should be specifically assigned for initiating the review, assigning the team, maintaining records of the team's findings, monitoring follow-up, and approving startup.

5.4 Recommendations/Follow-up/Closure

A project safety review could result in a list of action items as follows:

- Exceptions to standards and regulations,
- Recommendations of the review team based on knowledge and experience
- Issues requiring further study

Follow-up and closure of all action items, regardless of which type of project review has been conducted, are as important as conducting the review itself. Action items which must be addressed prior to startup should be clearly identified. The auditor should verify that the review procedure includes a provision for ensuring that all follow-up has been completed and documented prior to start-up. Action items must be addressed and the subsequent actions documented. This is true whether initial suggestions were followed, new ideas have been executed, or no action was necessary. The auditor should also confirm (e.g., through field checking) that all action items were addressed as specified in the action plan.

It is important to note that implementation programs will differ by company, site, or even department. What is of concern to the auditor is that a method is in place, is functional, and allows for documentation. Therefore, the ensuing paragraphs describe fundamental components of follow-up to project safety reviews without specific details. The auditor should be open to different implementation programs as long as they meet the necessary objectives.

5.4.1 Assigned Responsibility for Action Items

In order to ensure that follow-up is completed for each action item, responsibility for overseeing implementation must be assigned. The auditor should evaluate the process of assignment of responsibility to verify that the person assigned has the resources, authority, and knowledge to ensure the follow-up. This will help ensure that the people with the right expertise are addressing each finding, while one person is not being overwhelmed with a long list of action items.

5.4.2 Tracking System on Status of Action Items

An important component of the follow-up and closure period of project safety reviews is the tracking of action items. Without a formal tracking system, action items can be overlooked. There should be assigned responsibility for tracking the status of action items.

The auditor should be aware of the components which make a successful tracking system. These are listed below.

- Formal documentation
- Periodic reporting
- Updating ease
- Distribution of information

A list of action items with responsibility and status should be distributed to the project review team on a periodic basis. Periodic reporting of the action items is necessary to communicate their status or completion. Any change in status from the previous distribution should be noted and redistribution of the list should be made.

The purpose of such a tracking system is to ensure prompt attention to all safety concerns as well as to reassure the project review team that their action items (i.e., findings recommendations, questions, etc.) are being addressed. Periodic updating and distribution of the status report will help ensure prompt attention.

5.4.3 Resolution of Disagreements

Occasionally, there may be disagreement over a project safety review finding. The project safety review process should anticipate this situation and provide a mechanism for resolving disputes. The auditor should determine if such provisions exist, and whether they have ever been invoked (see Chapter 3).

5.4.4 Updating Process Safety Information

Process safety information updating is an often overlooked part of the project safety review and an auditor should be sensitive to this.

The auditor should verify that there is a clearly-defined initiating mechanism for updating the process safety information. One example is a checklist of process safety information which needs to be updated. This mechanism should recognize that the project safety review team itself will often disband at the end of the project.

5.4.5 Report

The project safety review report should contain a brief description of the process, the names of the people attending the review, the notes from the review, and a prioritized list of action items listing responsibilities and status of follow-up.

A final point of concern that is often overlooked in project safety review follow-up is a review of the changes that have been made as a result of the safety review. Significant changes to a process design may require additional safety reviews.

5.4.6 Dissemination of Findings

The practice of disseminating project safety review findings will vary from company to company. It is desirable to make findings available to other company facilities that have similar process units. The auditor should determine if the company has a policy on dissemination of project safety review reports, and whether it is followed.

5.4.7 Record Retention

A copy of the final project safety review report should become part of the process safety information documentation. The auditor should establish what provisions exist for retaining project safety review reports. The auditor should sample project safety review reports to test the system.

5.5 Summary

Project safety reviews are a critical element to ensure that major changes or additions to facilities and equipment are done safely. The auditor should be sensitive to the use of different methods for project safety reviews, but should assure that the method in use is comprehensive and is fully implemented.

6

Management of Change

6.1 Overview

Changes to processes are made for a variety of reasons, including but not limited to improved efficiency, operability, and safety. Changes can range from large facility expansions or new plants to minor changes in chemicals, technology, equipment, or procedures. Any change represents a deviation from the original design, fabrication, installation, or operation of a process. Even simple changes, if not properly managed, can result in catastrophic consequences.

Three types of changes should be managed at any location: technology, facility, and organization. Technology changes include modifications to operating procedures or parameters, use of new chemicals, and equipment modifications. Facility changes include modifications that involve substitution of equipment with "not-in-kind" replacements, either temporary or permanent. Organizational changes can range from substitution of new personnel to elimination of positions. These changes can have an impact on process safety if they result in insufficient staff or staff having insufficient skills or training, such that they hinder the management of process safety programs or result in slower or incorrect response to process upsets. Changes in technology or facilities may affect process safety directly, but they can readily be addressed through a management-of-change procedure. Organizational changes have an impact on the accountability and responsibility aspects of process safety and are therefore addressed in Chapter 3.

By far, the most challenging aspect of managing change is identifying that the proposed modification is in fact a change. When a change is identified, the next task is determining the level of review necessary prior to implementing the change. These are important elements of a good management-of-change procedure and they should be a primary focus of an audit.

This process safety management element interfaces greatly with other process safety management elements. The auditor should verify that the interfaces have been properly reviewed. This implies that the auditor should be familiar with the requirements of these elements to audit adequately if they have been implemented. The primary interfaces include

- Organization changes (Chapter 3)
- Procedure requirements (Chapter 3)

63

- Process safety information (Chapter 4)
- Project safety review (Chapter 5)
- Work authorization (Chapter 7)
- Preventive maintenance (Chapter 7)
- Process hazards review (Chapter 8)
- Training (Chapter 11)

6.2 Auditing Approach

A management of change audit can vary in scope and approach, based on company requirements. One approach might include the following steps (see Figure 6.1):

- Review written procedures and prior audit reports to verify adequacy of content and compliance with any procedures, standards, or guidelines for this element
- Identify recent facility and technology changes that should have followed the procedure
- Review completed change documentation to assure compliance with the written procedure
- Interview operations/maintenance personnel to verify that adequate training has been conducted prior to implementation of the change
- Identify changes to process safety information, operating procedures, and other documentation resulting from the change that may be included as part of the audits of those elements
- Verify that safety reviews are consistent with and follow appropriate process hazards analysis and pre-startup safety review procedures

The procedure should be discussed with the appropriate technical, operations, maintenance, and purchasing personnel responsible for implementation to fully understand how the system is supposed to work, and to ensure that the procedure addresses the key elements of management of change as discussed in Section 6.3. This interview session may involve only one or two facility staff with particular knowledge of the procedure and its applicability. A list of recent large and small changes may be identified during the discussion for later verification. The auditor should review the change reports on file to verify compliance with the written procedures and to ensure that all changes have followed the procedure. A random check to ensure that process safety information and other documentation have been updated should also be performed, or the required updates could be identified for follow-up when auditing those elements. Additional verifications may be done through interviews with both maintenance and operations personnel. This interview session may include a group of individuals representing a cross-section of the facility, including personnel from separate process areas. This effort should include verification that changes were properly reviewed and that training had been completed prior to implementation of the change.

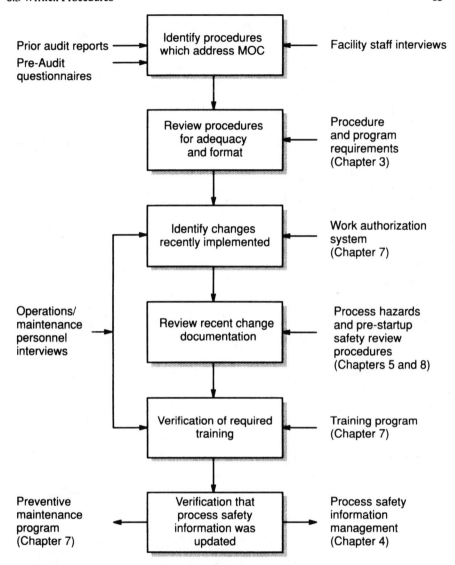

FIGURE 6–1. **Management of Change Audit Process**

6.3 Written Procedures

6.3.1 General Requirements

A written procedure for change authorization that follows the procedures (see Chapter 3) should be available. This procedure should address such items as support documentation requirements, format, review frequency, and authorization/approval requirements and distribution, along with the issues discussed below. The auditor should review the management of change procedure for conformance to any corporate or facility standards or guidelines for this element. One comprehensive procedure that addresses both technology and facility changes is preferable (see Figure 6.2). However, many procedures covering different types of technology and facility changes may exist.

Locations may address facility, experimental and instrumentation changes separately. The auditor should identify and review procedures that address management of change. If the management of change procedure requires an internal audit of the program, documentation supporting the completion of these audits and resolution and closeout of findings should also be verified by the auditor. If a variance to the procedure is in effect, the auditor should verify that the appropriate authorization and timetable for implementation has been given according to the variance procedures discussed in Chapter 3.

6.3.2 Definition of Change

The procedure should clearly specify what constitutes a change. One definition is to specify that all hardware changes must follow the procedure except "replacements in kind." A good definition of a procedural change is anything that requires modification of the process safety information. It is helpful to list certain types of changes specifically to ensure that they are identified as "changes" by facility personnel, and that there is no misunderstanding as to which changes must follow the review procedure. Some typical examples are listed below.

- Alternate chemistry
- New raw materials or additives
- Operation outside established safe operating limits
- Changes in operating procedures or parameters (temperature, pressure, charge quantities, order of addition)
- Experimentation
- New equipment or instrumentation
- Changes in area electrical classification
- Changes to computer software
- Changes to alarm, interlock or relief setpoints
- Bypass of alarms, interlocks or relief systems
- Equipment bypasses
- Equipment modifications
- Different materials of construction
- Temporary connections or equipment, including rentals and experimental equipment
- Decommissioning
- Job assignments

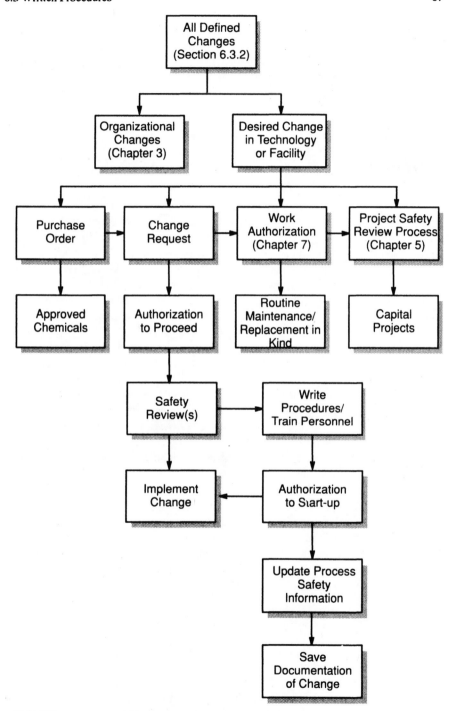

FIGURE 6–2. **Typical Management of Change Procedure**

When there is more than one procedure to address different types of changes, the auditor should verify that the scope of each procedure has sufficient overlap to include all technology and facility changes. Often when facilities rely on multiple procedures, they are too specific to address all possible changes and can result in changes being overlooked.

6.3.3 Identification of Change

The management of change procedures should have a clear mechanism for initiation including changes made for emergency reasons. To be effective, the mechanism should require documentation of the change in writing. One approach can be through a work authorization system in which any change is requested by a work order. This system can readily identify changes to facility and capital projects (which should follow the project safety review process) from routine or preventive maintenance work. A cross-check by a maintenance planner can provide additional assurance that all changes have been identified. Essentially all facility changes can be identified in this manner.

If changes are requested via a work authorization system, the author should verify that the required information is documented. The procedure should specify who is responsible for obtaining the proper review and authorization. Technology changes are generally easier to identify because they typically require coordination between technical, engineering, operations, and maintenance personnel. When new chemicals are involved, the purchasing function may provide a cross-check to ensure that the change has been properly reviewed.

The auditor should have a clear understanding of how changes can be made in the facility and ensure that all require some form of written documentation and approval by personnel responsible for operation of that facility. The auditor should verify that appropriate facility personnel have been notified of the procedures for managing change and that appropriate training has been conducted. Training should include operations or maintenance personnel responsible for approving work orders, or personnel with authority to change specifications for new chemicals or equipment.

A means for cross-checking or for requesting additional approval by organizational personnel who will implement the change should also be in place to provide assurance that the change has been properly identified and has been routed through the review process.

To verify whether changes have been identified and managed, the auditor might interview staff to discuss some changes that have occurred recently, and sample appropriate documentation. Staff who were affected by a change should be identified for follow-up interviews regarding training prior to implementation of the change. Additional verification can be obtained by a cross check of work orders and purchase orders with change requests.

6.3.4 Description of Change

Any change should be adequately described, including sketches or drawings, to ensure that the scope is clear. The technical basis for the change should be discussed along with the necessary updates to the process safety information. The impact of the change on process safety should be described in sufficient detail to help define the required safety reviews.

When reviewing a change request, the auditor should obtain a clear understanding of which process safety information must be updated as a result of the change, especially the piping and instrument drawings, operating procedures, and established safe operating limits. The auditor should also verify that information describing the technical basis and impact on process safety is adequate.

6.3.5 Temporary Changes

Temporary changes must have a reasonable duration specified. A maximum duration should be specified in the procedure to avoid having temporary changes become permanent. Appropriate durations for temporary changes would fall in the order of days or weeks rather than months. If the temporary change has to go beyond the duration specified, the procedure should require re-evaluation and re-authorization. Once the time approved for the change expires, the procedure should include provisions for removing temporary modifications and ensuring that all changes have been returned to normal condition.

The auditor should verify whether temporary changes have expiration dates and that changes are not left in effect beyond that date. If temporary changes were made permanent, the auditor should verify that they have been properly authorized. The auditor should also verify that expired temporary changes have been removed and that any changes in instrument setpoints, alarms, and interlocks have been returned to normal conditions.

6.3.6 Authorization

The person(s) who can authorize a change should be clearly identified in the procedure. If different levels of authorization are provided, a means to identify which changes require higher levels of authorization should be clearly described. The level of authorization should be related to the potential risks rather than to the level of investment or duration of the change. If alternates are allowed to approve changes, they should also be clearly identified, and they should comply with the variance procedures discussed in Chapter 3. The required level of safety review should be specified prior to authorization of the change, as well as any further authorization which may be required prior to startup.

The auditor should verify that the proper level of authorization was obtained for the type of change requested. In particular, changes that are implemented on off-shifts, weekends, or holidays should be checked for proper authorization. Also reauthorization of temporary changes that have been extended beyond their original expiration date, or those that have become permanent, should be verified by the auditor. The auditor should also evaluate whether the level of authorization is appropriate for the type of change. The auditor should recognize that changes with possible impact on more than one area or process should be authorized by someone responsible for all potentially affected areas or by someone from each area.

The auditor should try to identify recent changes that may have bypassed the system. He should explore the authority given to operating personnel to make changes to control, alarm, and interlock setpoints and to deviate from standard operating procedures. The auditor should evaluate whether the operators are exceeding their authority by operating at times outside established safe operating limits, or by changing/disabling alarm or interlock setpoints, particularly those accessible via computer.

It is also important to have a final authorization of the change prior to implementation. This should be incorporated into the pre-startup safety review procedure, but it could also be part of the change request. Other required sign-offs on the change request should also be verified.

6.3.7 Safety Review

All changes should have some level of safety review. Process hazards analysis techniques (discussed in Chapter 8) should be considered for the review based on an appropriate assessment of risk. A pre-startup safety review should be conducted according to the criteria presented in Chapter 5. The review should be conducted by a team representing a cross-section of expertise.

The auditor should verify that the procedure requires a safety review of the change prior to implementation, that there is a mechanism for assessing risk, and that the level of safety review is based on this risk assessment. A hazard analysis should be conducted on any changes which require engineering. She should verify that the safety reviews were conducted in accordance with the specified safety review procedures. The auditor should also include a sampling of the safety review documentation to verify completion and adequacy of review, including appropriate staffing. She should verify that action items required for completion prior to startup had indeed been completed. For changes determined to have a low risk potential, formal and systematic process hazards analysis procedures may not be necessary; a less rigorous review process may be appropriate.

6.3.8 Training

Prior to implementing a proposed change, training must be provided to all employees who are affected by the change. Any changes in safety, operating, maintenance, and emergency procedures should be included in the training.

The auditor should interview operations and maintenance (including contractors) staff who may have required training as a result of recent changes identified in Section 6.3.3. The auditor should ask the affected staff whether they felt adequately trained to operate and/or maintain the equipment affected by the change. The auditor should also verify that the training had been completed prior to implementation of the change. This verification should cover different shifts required to operate the equipment, not just those expected to start it up. As a separate, but complementary activity, the auditor should verify that any training was done in accordance with the site training program requirements.

6.4 Documentation

The requirements defined in Sections 6.3.4–6.3.8 should be clearly documented. A management of change form is typically used by the originator to document the key requirements and authorizations necessary to implement the change. Separate documents for the safety review findings and how they were addressed should also be kept on file. Before the change file is closed, there should be verification that the process safety information has been updated. The change file, including change request forms and safety reviews, should be retained in accordance with the procedure.

The auditor should verify that documentation of the change including a description, duration, authorizations, safety reviews, and training is complete. In addition, verification that the required process safety information had been updated should be checked, or a list developed for later verification when the process safety knowledge element is audited. Primary information to check includes the piping and instrument diagrams, operating and emergency procedures and safe operating limits. If the change required new equipment to be included in the site's preventive maintenance program, or required a change in preventive maintenance requirements or frequency for a piece of equipment, the appropriate maintenance records and procedures should be sampled.

6.5 Summary

Management of change is a challenging element to audit because of the broad range of changes that have to be included in a complete program and the interface of this element with many other process safety management elements. A key step is to identify all procedures that address management of change to allow sufficient time to review all of them in detail. When conducting an audit of this element it is important to have a good understanding of the processes at the facility and the procedures for addressing the other process safety management elements. A considerable amount of overlap and cross-checking of information is possible because of these interfaces, and these checks can uncover inconsistencies or gaps in other elements as well as those related to management of change.

7

Process Equipment Integrity

7.1 Overview

Assuring the integrity of process equipment must start with equipment design and continue through its fabrication, installation, and operation. Since the responsibility for these activities generally lies with the engineering, construction, and maintenance departments of a facility, the auditor should interview personnel from those departments. On the other hand, a large number of facilities rely on staff from their corporate engineering and construction departments, or outside contractors, for these services. If this is the case, the auditor should include interviews with individuals in those organizations as part of the audit.

In preparing for the audit, the auditor should identify the key individuals in the appropriate facility, corporate and contractor organizations. These individuals should have the following documentation available for review by the auditor:

- Design standards, specifications and records
- Fabrication specifications and records
- Installation procedures and records
- Inspection procedures and records
- Maintenance procedures and records

In addition, information from audits of other process safety management elements may be useful (see Figure 7-1), for example,

- Design standards and equipment specifications (from Chapter 4, Process Safety Knowledge)
- Specification of fabrication/installation tests and inspections (from Chapter 5, Project Safety Reviews)
- Critical equipment identification (from Chapter 5, Project Safety Reviews)
- Process changes which were initiated because of equipment failures (from Chapter 6, Management of Change)
- Review of incidents to develop the preventive maintenance program (from Chapter 9 Incident Investigation)
- Training of maintenance personnel and inspectors (from Chapter 11, Training)

The audit of process equipment integrity should include the following steps:

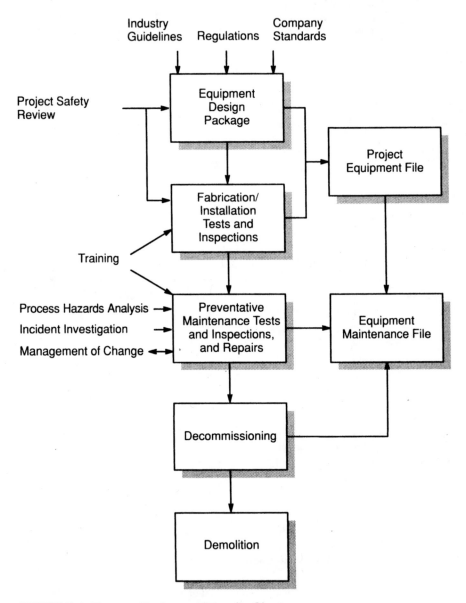

FIGURE 7–1. **Process Equipment Integrity Chart**

- Review of standards and procedures for format and content (see Section 3.2.2)
- Interviews with operations and maintenance department personnel to understand the PSM systems in place
- Interviews with construction/maintenance personnel (including contractors) to verify qualifications

- Sampling of equipment records for completeness and compliance with procedures
- Observation of new construction and maintenance work in progress for compliance with procedures

The auditor should have a clear understanding of the corporate and facility policies that specify what equipment must be included in the process equipment integrity program. The criteria for deciding what equipment should require a process integrity program should be based on how critical that piece of equipment is to ensuring the safety of the process. Most regulations and industry guidelines specify the processes to which they apply in terms of the chemicals handled in those processes. For those processes, the auditor should verify that critical equipment has been included in the process equipment integrity program. Some examples are

- Vessels and tanks
- Pumps and compressors
- Heat exchangers
- Piping systems
- Flexible connections (e.g., hoses, bellows, expansion joints or loading arms)
- Relief and vent systems
- Check valves and other backflow-prevention devices
- Emergency shutdown systems
- Controls, alarms, and interlocks
- Electrical distribution and other utilities
- Detection systems for flammable gas, toxic gas, and fire
- Fixed fire protection systems
- Grounding/bonding and cathodic protection systems

7.2 New Equipment Design, Fabrication, and Installation

Information detailing the design codes and specifications for new equipment should be contained in the design package. This package should include equipment specifications, data sheets, calculations, and detailed equipment drawings. The existence and completeness of this information should be audited as part of the process safety knowledge element (see Chapter 4).

The auditor should verify that the required tests and inspections, and acceptable limits for new equipment have been included in the design package. The auditor should also confirm that fabrication and installation recommendations from the project safety reviews (see Chapter 5) have been addressed. This information would typically be made available by the project engineering staff or manager.

For fabrication of new equipment the auditor should confirm that there is a quality assurance program that meets the project requirements. Contractors responsible for fabrication or installation may develop these criteria. The auditor should also verify that where applicable the following activities are specified prior to start of fabrication:

- Welding, fabrication, and non-destructive examination requirements
- Required approvals
- Qualification of employees (e.g., welder certification)
- Hold and witness points
- Quality assurance audits

Examples of quality assurance tests and inspections which may be conducted during fabrication include

- Radiographic examination of welds
- Mill tests for materials of construction
- Stress relieving
- Impact (Charpy V notch) tests
- Hydrostatic testing or pneumatic strength tests
- Nondestructive examination (NDE)
- Verification of dimensions and tolerances

The auditor should interview the personnel responsible for reviewing or participating in quality assurance programs and ensure, by sampling equipment records, that tests and inspections have been performed as specified in the requirements. Additional tests and inspections may be required during installation of equipment. Most of these requirements should be part of the design package or established with the installation contractor. Where manufacturers of critical equipment have specific installation instructions, the auditor should ensure that there is a mechanism for ensuring that they have been considered.

The range of tests and inspections that may be conducted during installation include

- Soil compaction
- Protection against frost heaves
- Strength of concrete
- Structural steel integrity
- Radiographic examination of field welds
- Bolting techniques
- Materials of construction for piping, gaskets, and packing
- Integrity of coatings/linings/refractory
- Leak testing of connections
- Testing of relief devices
- Function testing of instrumentation and controls, fire and gas detection systems, and emergency shutdown systems
- Proof testing of cathodic protection and grounding systems
- Electrical load testing of emergency equipment

The auditor should verify that fabrication and installation records include the date of the test/inspection, the name of the person who conducted the test, the test results and their acceptability.

Test/inspection records should be filed to facilitate retrieval. The auditor should verify there is a system in place to forward a copy of these records to the maintenance organization before startup of the equipment (see Chapter 5).

7.3 Preventive Maintenance

Critical equipment requires a preventive maintenance program to prevent or identify defects and minor failures before they can develop into more serious failures.

The auditor should verify that there is a mechanism for identifying critical equipment. He should also verify that any required tests and inspections, and their frequency and acceptable limits have been specified; and that there are written procedures for conducting each test or inspection. The auditor should also assure that there is a process for setting preventive maintenance frequencies. One method is to start with a conservative frequency and change inspection intervals based on test results. The manufacturer's recommendations for preventive maintenance should be a starting point unless there is experience on similar equipment. Incident history should also influence the need for, or frequency of, preventive maintenance. Some examples of preventive maintenance for equipment that may be critical to the process are

- Test and reset pressure relief valves
- Inspect and clean flame arresters and conservation vents
- Replace pump/compressor seals
- Inspect vessels and tanks
- Measure wall thickness or pressure-test piping and heat-exchanger tubes
- Proof test detectors, controls, alarms, interlocks and shutdown systems
- Function test emergency isolation and vent valves
- Replace rupture discs
- Replace or test hoses and other flexible connections
- Test/inspect emergency equipment and fixed fire protection equipment
- Analyze pumps and compressors for vibration
- Conduct thermographic analysis of electrical equipment

The auditor should verify that there is a system in place to analyze the maintenance history of equipment for trends, so that changes in the procedures, techniques, or frequency of testing, or the design of a piece of equipment can be made (e.g., a line that is found to require replacement every few years should be considered for a change of materials of construction rather than increasing the inspection frequency).

The auditor should then sample equipment records to verify that the appropriate tests/inspections have been carried out at the specified frequency and that the results fall within acceptable limits. If the equipment does not pass a required test, the auditor should verify that this information is communicated to the appropriate personnel, and that any repairs deemed necessary were made.

The auditor should also verify that there is a training program in place for maintenance personnel who conduct preventive maintenance (see Chapter 11).

7.4 Maintenance Procedures

7.4.1 Work Authorization

Process-related maintenance activities should require a written authorization. This authorization is particularly important to ensure that process changes are identified (see Chapter 6). It is equally important that maintenance work on equipment be documented to provide a record that can be analyzed as part of the reliability program.

The auditor should verify that process-related maintenance work requires a written description of the work and appropriate authorization. He should also verify that there is a mechanism for the quality assurance of materials and spare parts used for repair.

For critical service, or in instances where special materials of construction are required, there should be a method for verification of materials of construction. The pressure and temperature rating of equipment should also be confirmed by the maintenance coordinator or contractor personnel responsible for preparing material lists for maintenance jobs. Spare equipment, particularly portable or temporary equipment, should be inspected prior to use to ensure that it meets the design standards for the system in which it is to be used. If equipment or spare parts that are not replacements in kind are used, this would constitute a change which would require a safety review under the Management of Change element (Chapter 6).

Documentation of repairs to equipment should be retrievable. This documentation may include individual work orders authorizing the repairs.

The auditor should verify that there is a system to ensure that safety-related work will be given priority over routine maintenance. Maintenance procedures should specify that when work is completed, the job should be inspected to ensure that the equipment is safe to start up. This inspection should include verification that all valves are in the proper position and all associated alarms and interlocks are active.

7.4.2 Safe Work Practices

A critical part of any maintenance or project work on existing equipment is a set of safe work practices for preparing equipment for maintenance or tie-ins. Some examples of safe work practices are

- Hot work
- Line break
- Confined space entry
- Lockout/isolation

- Identification of underground process lines or utilities prior to excavation
- Lifting over active equipment or piping

The auditor should verify that safe work practices exist to control facility process hazards. Often permits or authorization forms are used as documentation that required tests and inspections have been performed. These would typically include testing for flammable or toxic vapors and oxygen concentration. Such permits should only be valid for a specified time, typically a work shift.

The auditor should identify maintenance or project work in progress that requires compliance with safe work practices, and verify that these procedures are being followed and that permits are complete and current. In addition, the auditor should sample maintenance records to verify that work authorizations and safe work permits have been completed as required.

7.5 Contractors

The use of contractors for equipment maintenance is common. A study has been completed on the use of contract labor in the U.S. petrochemical industry (John Gray Institute, 1991). This study found that contract labor is used to conduct approximately one-third of routine maintenance and one-half of the maintenance required during turnarounds. Reliance on contractors makes it important that contractors are aware of hazards in the workplace and comply with each facility's safe work practices. A facility may use a number of contractors, some for routine maintenance and others for special tasks. The auditor should verify that the safety record of contractors is considered prior to award of contracts. The auditor should also verify that a system exists for contractor orientation and that it includes a review of the emergency plan, hazard communication, and the facility's safe work practices. The auditor should verify there is a system for ensuring that the contractor has trained his employees in the procedures necessary to work safely at the facility. If the auditor has the opportunity to observe maintenance or tie-in work (see Section 7.4), she should also verify that contractors are following the facility's safe work practices.

7.6 Decommissioning and Demolition

When a piece of equipment is no longer needed for operation, it should be decommissioned. Unused equipment can be a potential hazard if not isolated and/or cleaned properly.

The auditor should verify that there are procedures in place that specify the requirements for isolation and cleaning of equipment to be taken out of service. She should evaluate how equipment recently taken out of service was handled, and determine if special procedures for cleaning and disposal of hazardous residues or materials have been written (e.g., disposal of asbestos). The auditor should verify that if equipment is subsequently recommissioned it has had the appropriate safety reviews (see Chapter 6).

When equipment has served its useful life, it should be dismantled, removed, and/or demolished.

7.7 Summary

Process equipment integrity must be considered in all phases of equipment life from its initial design, through fabrication, installation and operation until it is demolished. The audit of this element must include quality assurance systems during fabrication, installation, and repair of equipment as well as systems for preventive maintenance during the useful life of the equipment.

8

Process Risk Management

8.1 Overview

Process risk management involves the identification, evaluation, and control of potential hazards that may be associated with existing operations, modifications, new projects, acquisitions, toll processors, and even customer/supplier activities.

Some corporations, divisions within corporations, or individual facilities have developed formal procedures or guidelines on process risk management and/or its underlying technical studies. These guidelines may specify the methods for the technical studies, the decision-making process, the scope of the technical studies, and the reporting format. The auditor must become familiar with these guidelines and use them as the primary standard for auditing the activities that have been undertaken for the risk management of a facility.

The hazardous conditions that are typically addressed by these guidelines include fires, explosions, and/or releases having acute toxic effects. Analyses then consider basic causes, such as postulated equipment failures, human error, or lack of administrative controls (e.g., PSM systems) that contribute to events that can lead to these hazardous conditions. The risk of these events is often expressed in terms of the severity of their adverse effects and the likelihood of their occurrence. For risk management decisions, the typical adverse effects (consequences of concern) include impact on facility personnel, the public, the environment, or the business.

The key components of a process risk management program generally address:

- Hazard identification
- Risk assessment of operations
- Reduction of risk
- Residual risk management
- Process management during emergencies (see Chapter 12)
- Encouraging client and supplier companies to adopt similar risk management practices
- Selection of businesses with acceptable risk

The basic concepts and purposes of each of these components have been discussed in the *Guidelines for Technical Management of Chemical Process Safety.*

An audit of a process risk management system should begin with a review of the elements of the facility management system that deal with process risk management.

These elements may be found in one cohesive package, or they may be found in a number of documents (e.g., safety standards, and engineering guidelines). From interviews with key members of plant management, the auditor should understand the risk management goals, how the acceptable level of residual risk is determined, special requirements and practices for process risk management, risk management programs that are in place, and risk assessment and other studies that have already been completed. Also, the auditor should establish the existence of any guidance documents, specific guidelines (e.g., CCPS *Guidelines for Technical Management of Chemical Process Safety*), and regulatory requirements that may influence or form the risk management program, as well as any technical studies that have been completed. The auditor should also try to identify several decisions, projects, or modifications that should have been included in the risk management program. He can then use them in the audit to verify whether the guidelines have been followed.

An audit of process risk management should cover the following:

- Thoroughness and cohesiveness of risk management programs and guidelines, including staff awareness of these programs
- Technical studies, including hazard identification and risk assessment studies of operations, modifications, and new projects, in terms of appropriate use, documentation, and approval of such studies
- Risk reduction measures based on these studies, as well as decisions made without formal studies
- Management plan for residual risks
- Policies regarding risk management practices of customers and suppliers
- Risk management procedures for new business acquisitions

These areas are discussed in some detail in the following sections.

8.2 Hazard Identification

The objective of hazard identification is to review an engineered system as thoroughly as is reasonably practical, to identify deviations of the system from the design intent that could lead to an event with the potential for causing undesired consequences. As such, hazard identification is the foundation of a solid risk management program and should be conducted at numerous stages in the life cycle of a process. Project safety reviews (see Chapter 5) are the primary mechanism for conducting such studies on new projects. Process hazard analyses are the primary form of hazard identification studies for existing facilities, while management of change (see Chapter 6) procedures address changes to these facilities. The emphasis of the audit of process risk management should be on evaluating the system for conducting hazard identification.

The auditor should identify standards and guidelines that may exist for hazard identification. Also, the auditor should interview plant management to determine their understanding of the study objectives, and their role in the documentation and acceptance process. The auditor should then review a sample of hazard identification reports and interview the key staff involved in producing the reports to verify that the guidelines have been followed or verify that others are reviewing these documents as part of project safety reviews or management of change audits, and determine if the study objectives were met. Specific items for review are discussed in the following sections.

8.2.1 Scope of the Study

The usefulness of the results of a hazard identification study is dependent upon the initial decisions regarding the scope of the study. The scope is often defined in terms of level of detail of study, consequences of concern, and system boundaries. To audit the elements of the scope of the study, the auditor needs to identify the standards or guidelines that specify or address these elements. For example, there may be corporate guidelines that require any systems that handle chlorine or hydrogen chloride to undergo a periodic hazard identification study, while systems handling less hazardous materials may need review only when they are used in large quantities or above their boiling points. Guidelines may also specify whether the impact of concern is for example: any release, a vapor cloud release, a release above a specified quantity, or a release with offsite potential.

The level of detail includes such considerations as looking at individual pieces of equipment versus subsystems, and looking at internal versus external causes of failures. There are two general categories of causes of failures: (1) direct failures of the equipment used to process the chemicals (e.g., a gasket leak in a chlorine system), (2) and failures unrelated to the system that can lead to a release from the system (e.g., a rupture of a line during excavation or because of a crane accident). Failures in the latter category are also known as external events.

Risk management programs also address specific undesired consequences. These may be

- Fatalities or injuries to employees and/or the public
- Property damage and business interruption, or
- Environmental damage

The undesired consequences of concern are selected per the objectives of the specific study and the overall risk management program. If the objective is to minimize public exposures, the study is usually limited to those release scenarios that lead to adverse effects outside a facility's fence line.

System boundaries should be clearly defined in terms of geographical, systemic, and functional limits. The auditor needs to examine whether the boundaries have been clearly stated in selected hazard identification reports and whether these boundaries are consistent with the guidelines.

8.2.2 Methodology Selection

Several methods are available for identifying the hazards of an operation. Some of the common approaches identified in the *Guidelines for Hazard Evaluation Procedures : Second Edition with Worked Examples* include

- Hazard and Operability Analysis (HAZOP)
- Failure Mode and Effects Analysis (FMEA)
- What-If Analysis
- Checklist
- What-If/Checklist Analysis

The book referenced above provides guidelines for selecting the appropriate methodology for hazard identification. Some corporations also have special guidelines for this purpose. The selected methodology should be consistent with regulatory requirements, as well as company guidelines. If there are no corporate or site guidelines on selecting a methodology, the basis for selecting the methodology should be documented; the auditor should interview those conducting such studies to determine how they select a methodology.

8.2.3 Implementation Practices

Hazard identification studies should be led by an individual knowledgeable in the selected methodology. It is imperative that the hazard identification team include personnel who are knowledgeable of the operational, engineering, and maintenance aspects of the system. The auditor should verify that team leaders have experience in applying the selected techniques, and that at least one team member is knowledgeable in the process being reviewed. The responsibility for assuring that such studies are conducted should also be clearly defined.

Results of the study should be documented in a written report. The report format can vary significantly, depending on the methodology used, and hazard identification may be addressed as part of a more detailed hazard analysis report. Reporting can be as simple as presenting a list of the findings and recommended modifications, or it can include detailed discussions of the methodology and the specific hazard scenarios. The level of detail in a report should be consistent with the objectives of the study and external or company requirements. For example, some regulations may require certain reporting practices. The auditor should determine if such guidelines exist, and review reports to see if they have been followed. The auditor should also verify that appropriate individuals receive copies of these reports.

Hazard identification should be repeated on a periodic basis. The process risk management program should specify the frequency of these studies, which may be based on regulatory requirements, corporate practices, the outcome of previous studies, changes in the process (see Chapter 6), or a combination of these factors. The auditor should determine if there is an established set of guidelines regarding frequency of hazard identification, and verify adherence to the set frequency.

8.2.4 Study Recommendations

A hazard identification study generally results in a set of recommendations for modifications to equipment, operating procedures, or administrative controls. The auditor should determine if such recommendations have been documented, and assess whether a management system exists to track the status of the action plan developed to address the recommendations. Actions taken to address recommendations should be documented, including any decision not to take action and the justification for that decision. The auditor should determine the existence of this paper trail, and also verify that recommendations were closed out within the specified time.

8.3 Risk Assessment of Operations

A risk assessment entails a detailed look at the hazards identified in terms of both their potential consequences and the likelihood of their occurrence. A hazard identification study is not always sufficient to support risk management decisions. If the recommended modifications are extensive or costly, a better measure of risk may be needed to determine the most effective risk reduction measure. In such cases, further analysis is often warranted.

The auditor should identify any corporate, site, or regulatory guidelines that may exist for risk assessment. The auditor should review a number of risk assessment reports, and interview key staff involved in producing the reports to verify that the guidelines have been followed and are readily available and known. Also, the auditor should interview plant management on their understanding of the objectives of a particular study and whether the objectives have been achieved, as well as their role in the documentation and acceptance process. The auditor may also interview the corporate staff if they supplied the guidelines or expertise for a particular study.

8.3.1 Scope of the Study

Most risk assessments are qualitative or semi-quantitative in approach. A typical objective of a qualitative risk assessment is to provide a ranking or prioritization of the individual release scenarios. In some instances, a quantitative risk assessment may be used to estimate the risk level of the process, which may then be compared with other risks, or to company guidelines, or used to determine how much risk reduction is necessary.

Risk assessment studies may be conducted on an entire process or unit, or only on selected hazard scenarios. They may be geared toward offsite fatalities, environmental damage, or other consequences. A risk assessment report should discuss the objectives, the consequences, the equipment/operations covered, the selection of methodology and the level of detail (i.e., screening versus detailed, qualitative versus quantitative, or conservative versus best estimate); the auditor should interview staff and review several reports to verify that such issues have been included in the reports.

QUALITATIVE RISK ASSESSMENT

The scoping of qualitative risk assessments requires engineering judgement. The auditor should verify that there are guidelines for qualitative risk assessments that address

- The inclusion or exclusion of utilities and other site-wide services
- The treatment of secondary or "domino" events
- The assumptions used to develop frequency and consequence estimates
- The definitions of qualitative estimates of risk (e.g., high, medium, or low)
- The basis for risk prioritization
- The conditions under which additional analysis should be conducted

The first two of these items are also important in a quantitative risk assessment, but they are more likely to be overlooked in a qualitative study.

QUANTITATIVE RISK ASSESSMENT

For quantitative risk assessments one of the most critical aspects of the scoping process is the consistency of assumptions within a study and between studies. The auditor should verify that guidelines or standard assumptions exist for:

- The impact level of concern (serious injury or fatality) in terms of specific modelling criteria for overpressure or thermal radiation or toxic exposure
- The probability of suffering a specified consequence, given exposure to a certain level of hazard
- Selection of consequence modelling packages/models and input parameters (i.e., one set of meteorological conditions or a range of conditions)
- Presentation of results, for example, F-N curves (graph displaying chances of exceeding various numbers of consequences), risk contours (individual risk levels around a site), or expected number of fatalities
- Typical data bases from which failure rates should be obtained, or use of standard/generic failure rates
- Treatment of uncertainty
- Treatment of population—uniform density versus actual locations or average versus day and night

The auditor should verify the existence of and adherence to such guidelines by reviewing quantitative risk assessment reports and interviewing staff. Corporate guidelines should include regulatory requirements.

8.3.2 Methodology Selection

Several methods for risk assessment are described in the CCPS book *Guidelines for Chemical Process Quantitative Risk Assessment*. The main difference among the methods is in the level of detail. The appropriate level of detail is usually based on the materials being handled, the complexity of the process, estimated risk level, and the available resources. For example, if the system design includes safety systems such that simultaneous failures need to occur to produce a major release, or if an estimate of the actual risk levels faced by the public is needed rather than just prioritizing the hazard scenarios, then quantitative risk assessment may be useful. The auditor should verify that documentation exists regarding the basis for the selection of the methodology used.

Corporate policies or regulatory requirements may specify the methodology, level of detail or format of the results for the risk assessment. For example, California regulations (California, 1989) require the handlers of acutely hazardous materials to identify "the most likely" hazard and conduct a plume dispersion analysis. In this case a semiquantitative assessment is performed using those release scenarios that are deemed to qualify as "the most likely" hazard. As another example, there could be guidelines that specify when to use a specific consequence analysis or dispersion model. The auditor should try to determine the existence of such guidelines, and then review the adherence to them.

8.3.3 Implementation Practices

Personnel involved in hazard identification may also conduct hazard evaluations or qualitative risk assessments. However, the skills and experience involved in conducting quantitative risk assessments are quite different from those needed for hazard identification. Therefore, the expertise and knowledge of the personnel conducting risk assessments are often quite different from those involved in hazard identification studies. For example, depending on the scope of work, the risk assessment team may require personnel who are experienced in fault tree analysis, probabilistic analysis, meteorology, source term estimation, and plume dispersion analysis. Often quantitative risk assessments are conducted by consultants or corporate staff. Therefore, it is imperative that plant personnel review the assessment results to assure that process conditions and other assumptions that are used in the analysis reflect current operating practice.

Documentation standards or guidelines should be available, particularly for studies performed by consultants. The reports should include objectives, scope, discussion of methodology and assumptions, results and recommendations. The auditor should review reports to determine consistency in content.

Residual risk management (see Section 8.5) requires periodic review and reanalysis of the technical studies, as does management of change (see Chapter 6). The corporate guidelines for residual risk management may specify how often such reanalyses must be conducted, as well as under what circumstances a reanalysis should be initiated ahead of schedule. The auditor should verify the existence of such guidelines and determine the adherence to those guidelines by reviewing older studies and following their paper trail.

8.3.4 Study Recommendations

Risk assessments of operations, as for hazard identification studies, often lead to a list of recommendations for risk reduction measures (see Section 8.4). These recommendations, their priority and implementation status, and the documentation of any actions (or nonactions) should be addressed, as was discussed for hazard identification studies (see Section 8.2.4).

8.4 Risk Reduction Activities

The outcome of hazard identification and risk assessment studies can take the form of a series of risk reduction measures, a prioritization of measures, and an action plan and schedule for implementing these measures. The auditor should determine if the facility staff have evaluated the effectiveness of the measures, and reviewed the potential added risks that may result from implementing the measures. (Some of this information may be found in the original hazard identification or risk assessment study.)

There are many methods for identifying risk reducing measures, including technical studies (see the preceding two sections), review of the systems against special safety standards (e.g., fire code regulations), management systems review for safety, and input from safety committees.

The auditor should review the risk reduction activities for the following elements:

- Corporate or regulatory risk acceptance guidelines
- The types of risk reducing measures considered
- Assurance that new risks are not introduced because of the risk reduction measures
- Dates and responsibilities for implementing risk reducing measures
- Effectiveness of the risk reducing measures
- The approvals of individuals with accountability for process safety

Means to establish acceptable risk can vary. The auditor should conduct interviews with appropriate decision-makers to determine how risk acceptability judgements are made to see if they are made consistently. For example, conformance with corporate engineering standards, adherence to industry standards, or avoidance of releases with offsite impact are each approaches to defining risk acceptability.

The risk reduction program should also allow different types of risk reducing measures. It should include changes in management systems, hardware, process parameters, process design, chemicals, safety systems, monitoring systems, release mitigation devices, and administrative controls, as appropriate. The auditor should review any statements about the guidelines to assure that none of these types of measures has been excluded. The auditor should verify through discussions with appropriate personnel that they feel free to suggest risk reduction measures and management addresses these suggestions.

The outcome of the decision process is a risk reduction plan that provides an implementation schedule for changes based on recommendations from the studies and the decisions of the corporate and facility management organizations. The risk reduction plan is key to process risk management and process safety management program development, and should contain clearly defined responsibilities.

The auditor should verify that responsibilities for implementation of the risk reduction plan have been defined and that the facility is meeting the schedule for its risk reduction plan. The auditor should also verify that if there are some recommendations from the technical studies that management has elected not to implement, that the reasons for not implementing these recommendations are documented.

8.5 Residual Risk Management

Risk reduction actions reduce risk levels, but they may not completely eliminate the risk from a particular hazard. Thus, there is almost always some degree of residual risk present after the risk reduction plan is fully implemented. Management of the residual risk is an important element of a process safety management program, and involves ongoing review and reconsideration of process safety-related conditions and controls.

Residual risk management should entail periodic reviews and revisions of risk assessment and hazard identification studies to confirm that the risk of the operations has not increased. The auditor should verify the existence of a paper trail for the periodic reviews, the results of the reviews, the subsequent action items, and the corresponding schedule for implementation. Guidelines should exist that define the frequency of reviews, a format for the documentation, and approval and communication channels for the findings. The auditor should probe for their existence through interviews with staff and reviews of documentation, and should verify that there is a tie-in to management of change procedures (see Chapter 6).

The technical studies should be based on a set of assumptions regarding the system design, plant layout and operating conditions that will include a degree of uncertainty. For example, the reactivity of a substance may not be accurately known at the time of the study. The auditor should verify that any periodic reviews revisit the underlying assumptions and areas of uncertainty to confirm that the changes in plant conditions and new information regarding the operation since the last review have been properly evaluated for their impact on the residual risk level.

New information may come from additional experience gained through ongoing facility operation. For example, near-miss incidents may lead to the discovery of accident scenarios that were not considered in an initial risk assessment. The continued plant operation also adds to the statistical database of equipment failure and human errors. Significant differences may exist between the actual (experienced) failure rate and that used initially in a risk assessment. Thus, the auditor should verify that the residual risk management program includes a periodic review of the assumptions used in the technical studies as may be defined in corporate guidelines.

New or improved consequence models may also become available, and they may provide better insight into the behavior of certain releases. More accurate toxicity data may become available, and may result in increased or decreased risk levels. Similarly, risk acceptability guidelines may be revised because of changes in corporate philosophy or public tolerance. The auditor should verify that the residual risk management process addresses these types of change, either as initiators of a review or as an element in a periodic review.

The auditor should verify that the following elements are considered in the residual risk management program:

- Periodic review of assumptions and methodology
- Preparation of written documents that describe the review of assumptions and methodology
- Communication of findings of periodic reviews to appropriate parts of the organization
- Identification and completion of follow up action items

8.6 Customer/Supplier Facilities and Practices

The auditor should determine if there is a system for defining high risk raw materials or products. For these materials, the corporation should assure that its customers and suppliers have risk management programs. Contractual or other less formal mechanisms can be created to assure that proper risk management practices are implemented. For example, the supplier, carrier, and customer companies may be asked to participate in inspection or audit programs.

The auditor should verify that guidelines exist for determining which materials are covered by this program, and for selecting or accepting suppliers, customers, and carriers/shippers based on their risk management practices or willingness to abide by the corporation's recommended practices. Also, the auditor should verify whether the guidelines specify recommended practices for customer unloading and storage or supplier-certified product analysis, as well as actions that facility managers should take if failures are noticed in implementation of the agreed-upon risk management measures.

The auditor should interview plant management to assess whether any contractual or other formal agreements exist between the plant and its suppliers, customers, and carriers. The auditor should verify that there are records of inspections of the facilities and procedures for the suppliers, carriers, and customers. Of particular interest should be the transport vessels that are owned by other corporations, but are used by the facility. The auditor should confirm that a program exists to review the risk management program of the carrier or customer corporations to assure the integrity of these vessels. The auditor should determine if an historical accident database maintained by the carriers/suppliers is used to help in carrier selection/rejection.

Where a facility uses contracted or toll operations, there should be an assessment of the relative risks. The auditor should request to see documentation of such considerations and the measures that are in place to assure that good risk management practices exist and are encouraged. Examples would be periodic inspections, hazard identification studies, or the extension of facility safety reviews or programs to cover these operations.

8.7 New Businesses

Understanding the process safety risks associated with a new venture or new acquisition is an important element of the overall corporate risk management policies. The process safety risks associated with a new business or acquisition should be clearly and accurately stated to the business managers responsible for the new venture. Also, after the purchase, the process risk management practices of the new facilities should be integrated with the corporate policies and practices.

The auditor should confirm the existence of guidelines and policies for assessing the process safety risks of a new business. The auditor should also interview key management to identify recent acquisitions and verify whether the corporate guidelines were used in the decision-making process and whether the acquired facility has incorporated the corporate hazard identification or risk assessment guidelines. The guidelines should specify the initial review methods and a format for presenting the findings to upper management. The guidelines may also address the methods and timing for bringing "unacceptable" operations up to corporate standards.

8.8 Summary

An audit of process risk management may require extensive reviews of documents and interviews with staff if the process risk management program is not consolidated and well-documented. However, most facilities and organizations should have reasonably comprehensive guidelines or standards. The auditor should not be unduly prejudiced by the organization or format of the guidelines or standards. One potential area of weakness of many programs, however, is the thoroughness with which they address all types of operations—existing facilities, modifications, new businesses, customers and suppliers. The auditor should keep the full range of operations in mind as she examines the process risk management program in place for existing operations to see if comparable items exist for the other operations.

9

Incident Investigation

9.1 Overview

"Those who cannot remember the past are condemned to repeat it," said George Santayana in his work *Life of Reason*. His aphorism aptly captures the essence of why incident investigation is needed. Systematizing the learning process, and capturing and applying the lessons learned is the role of the incident investigation system.

Since 1987, several organizations and agencies have developed comprehensive system requirements for managing process safety (see Chapter 1). While there may be some variation in the principal elements and emphasis, every guideline includes incident investigation as an element. This illustrates that incident investigation, analysis, and follow-up constitute an essential feature of any process safety management system, and safety cannot be properly managed without it.

There is no general template for incident investigation that is universally appropriate. To audit this element effectively, the auditor should have an appreciation of the objectives and methodology of incident investigation. Because there is no industry standard for incident investigation, auditors of this element should familiarize themselves with the philosophy and content of some sample incident investigation procedures. Examples of incident investigation procedures can be found in a number of references (*Guidelines for Investigating Chemical Process Incidents*).

There are three general categories of incidents:

- *Major accident:* an incident involving multiple injuries, a fatality, and/or extensive property damage
- *Accident:* an incident limited to a single injury and/or minor property damage
- *Near-miss:* an incident that has the potential for injury and/or property damage

Each of these categories of incidents should be reported and investigated, although the depth of investigation may vary with the severity of the incident.

When developing procedures to deal with incidents, an important distinction must be made between reporting and investigating. Taking a position that *all* incidents, regardless of their seriousness, should be reported, is practical. This can and should encompass near-miss incidents based on the definition of an incident. On the other hand, extensive investigation of all reported incidents falls beyond the resources of most companies. Consequently, guidelines for determining the level of investigation for an incident are essential.

An audit of the incident investigation management system should involve the following actions:

- Identify and interview key staff involved in implementation of the incident investigation system
- Review procedures for format and content (see Chapter 3)
- Identify any recent incidents that should have been reported and investigated
- Review and assess incident reports and investigations
- Interview personnel involved in recent incidents
- Assess the system for appointment of investigators
- Verify that recommendations have been addressed
- Verify that findings have been reviewed and disseminated
- Verify that incidents are being classified and analyzed for trends

Some areas of incident investigation where weaknesses are common include:

- *Incident reporting:* It is difficult to get personnel to report near-misses, especially those involving their own errors. Inadequate reporting generally results if personnel are reprimanded for their errors
- *Information:* The incident site is not preserved and data essential to an investigation are not gathered promptly
- *Investigation:* Analysis for *root causes*[1] is generally the exception rather than the rule. Consequently, recommendations tend to address symptoms rather than underlying causes.
- *Follow-up/closure:* The status of recommended action items is often not tracked

The auditor should also be aware of the interfaces of the incident investigation element with other management systems (see Figure 9-1) in particular:

- Any changes to prevent similar incidents (Chapter 6, Management of Change)
- Use of incident investigation reports in process hazards analyses (Chapter 8, Process Risk Management)
- Training of investigators (Chapter 11, Training)
- Notification of emergency personnel and facility management (Chapter 12, Emergency Response Planning)

---

1 *Root causes* are defined as management system failures, such as faulty design or inadequate training, that led to an unsafe act or condition that resulted in an incident; underlying cause. If the root causes were removed, the particular incident would not have occurred (CCPS, *Guidelines for Investigating Chemical Process Incidents*).

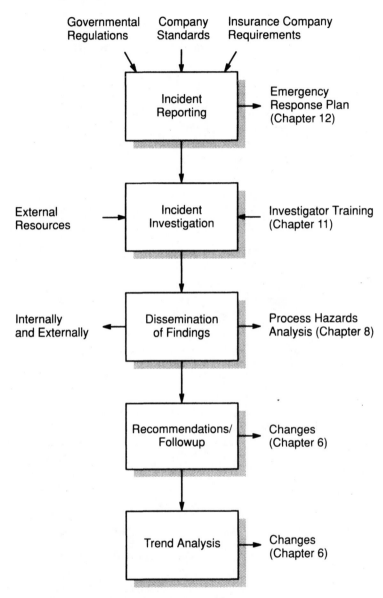

FIGURE 9–1. **Incident Investigation Flowchart**

9.2 Incident Investigation System

The first step in an incident investigation system audit is to interview the personnel responsible for coordination of the incident investigation system (i.e., the incident investigation coordinator) to gain an understanding of the management system that is in place. The next step is to obtain a copy of the incident investigation system description or at least the reporting form(s). This should be reviewed for format and content before beginning the data-gathering process.

The incident investigation system should address all three types of incidents discussed earlier, as well as the requirements and responsibilities for reporting, investigation, follow-up, record keeping, and trend analysis. The specific items to be audited are discussed in more detail in the following sections.

The auditor should establish whether the system complies with facility requirements (Chapter 3) for updating and approval.

9.3 Reporting Mechanism

9.3.1 Definition of Incidents

Incident investigation begins with incident reporting. Therefore, the definition of an incident that requires reporting should be reviewed to determine if it is clear and concise. The National Safety Council's (NSC) definition of an incident is "an unplanned, undesirable event that adversely affects the completion of a task. All accidents are incidents." A definition that is more applicable to the chemical industry is "any occurrence, condition, or action which did or could have resulted in personal injury, or damage to the plant, community, or the environment."

Major accidents and accidents can readily be identified as incidents; however, near-misses are more difficult to define and to communicate to facility personnel. Potential accidents or near-misses are included in both of the definitions given above. Incidents are often classified to assist in determining the extent of investigation and may require formal reporting to governmental agencies or insurance companies. Some examples are

- Fatalities
- Chemical releases
- Equipment damage or production loss above a minimum dollar value

The auditor should note whether such a classification has been defined, since the classification often determines the investigation requirements for the incident.

While clear definitions are important, it is equally important that facility personnel have a good understanding of what constitutes a reportable incident. This is the critical link in the investigation procedure; incidents that do not get reported cannot be investigated. For example, not reporting a near-miss involving an error of omission or commission may conceal a significant deficiency in the area of training.

The auditor should try to determine the type of incident that personnel generally understand requires reporting. This can be done through interviews. There is a trade-off between reporting minor deviations, which may overload the system such that adequate investigation of all incidents does not occur, and not reporting near misses. Some incident reports should be reviewed for proper reporting and investigation.

While the goal of incident investigation is to prevent recurrence of incidents, a company also needs to be prepared to deal with episodic events that could have significant impact on its facility and the corporation. Investigation of major accidents may involve special techniques and resources that a facility might not possess. Therefore, the auditor should determine whether there is a plan for investigating major accidents, particularly with respect to identification of, and means to access, corporate and outside resources (*Guidelines for Investigating Chemical Process Incidents*).

9.3.2 Initial Reporting

A standard format should be used to report incidents initially. The auditor should review the format to see if it is consistent with the written procedure. The incident reporting process should be designed to elicit the recording of factual information and should have space to list, at least, information needed to satisfy reporting requirements to governmental agencies, insurance companies, or corporate organizations. The auditor should review the process to verify that the required information is reported.

9.3.3 Responsibilities

The reporting chain starts with the employee. This may seem basic, but the auditor should verify that employees understand that they are responsible for reporting incidents. Timely reporting of incidents, particularly minor injuries and near-misses is also important. The auditor should verify that incidents have been reported within any time constraints set by company policy. If someone other than the person who observed the incident (e.g., the employee's supervisor) completes the incident report, both individuals should be identified on the report.

The auditor should verify that the system works as intended by interviewing key staff, such as the plant manager, incident investigation coordinator and members of the emergency response team.

9.4 Investigation

9.4.1 Criteria for Investigation

Incidents need to be investigated beyond initial reporting to derive maximum benefit. As a minimum, incidents should be investigated by a qualified investigator. The auditor should determine if there is a system for determining when team investigations are required and how they are to be organized. Typically an incident investigation should be initiated within 48 hours of its occurrence. The auditor should verify that investigations have been initiated within the required time limits and are completed.

Incidents requiring activation of an investigation team should be included in the system. For example, criteria for activating an investigating team might be

- Major accident
- Accident
- Near-miss with serious potential

9.4.2 Investigation Team

The auditor should be conscious of two key aspects of team composition, namely, objectivity and appropriate skills. Therefore, it is best if the team leader has no vested interest in the area in which the incident occurred. The auditor should verify that there is a system for selecting investigation team members based on qualifications.

The skills needed by members of an investigation team depend on the nature and extent of the incident. At least one member of the team should have first-hand knowledge of the process in which the incident occurred. Major incidents involving extensive site damage may call for a corporate representative or consultants, in addition to site personnel, to take part in the investigation. The auditor should assess whether any guidelines are provided that link incident type or class with specific representation. For example, a corporate risk management representative or consultant might be included for incidents involving

- Fatality
- Contractor incidents with liability implications
- Serious environmental or plant damage

The incident investigation system should define the responsibilities of the investigation team and to whom they are to report. Reviewing past reports is a method for determining whether teams have met their responsibilities. This should include verification that an analysis for root causes was conducted as part of the investigation.

The team should have the authority to interview personnel who can provide information pertaining to the incident and access to additional resources (such as technical or legal expertise) as needs are identified during the investigation. The auditor should verify that the investigation team has the means to access these resources.

9.4.3 Investigation Process

It is important that the investigation focus on gathering facts and preserving evidence, not on finding fault. Items that should be considered are

- The need to interview personnel quickly before memories fade
- The need to collect physical and process evidence before it is lost

- The need to think globally in securing evidence, since it is not always possible to anticipate in what direction the subsequent analysis will go
- The need for recording techniques such as audio and videotaping, photography, boroscope inspections, radiography
- The need (in some cases) to take samples of process fluids for subsequent analysis

An effective way to handle this aspect of the audit is to interview plant personnel who have participated in investigations. The auditor should talk to both investigators and witnesses to obtain a balanced perspective.

After examining the facts, the underlying or root causes of an incident need to be determined. The auditor should review past investigation reports to assess the methodology used and findings, and also to verify that the findings were incorporated into recommendations. The auditor should also verify that the investigator or team have had experience or training in root-cause analysis.

For major accidents, governmental agencies and insurance companies may conduct their own investigations and the law department may become involved. The auditor should determine whether the system has anticipated the need to coordinate investigation activities with outside organizations in terms of gathering facts and evidence, and the timing of reporting. If the site has ever had such an accident, the auditor should interview someone involved with the investigation to determine the effectiveness of coordination with external organizations.

9.5 Investigation Reporting

A written report of the investigation should be required. To meet this requirement, the auditor should review copies of several reports spanning the audit period, and verify that necessary information has been included in the report and that the investigation was completed within a reasonable time frame. The report should contain at least the following information:

- Date and description of the incident
- Factors contributing to the incident
- Recommendations resulting from the investigation
- The names of the people who conducted the investigation

The auditor should review reports to see if the findings and recommendations have been summarized and are readily apparent. She should review a sample of the recommendations to determine if they were based on a root-cause analysis.

She should also review the reports for findings/recommendations that are of a generic nature and discuss those findings with the facility staff to determine if the recommendations were reviewed for application to other units or plants within the company.

9.6 Dissemination of Findings

Identifying "lessons learned" is of little value if they are not conveyed to individuals and organizations who could experience the same incidents.

9.6.1 Internal Distribution

The auditor should review the incident investigation system to determine whether there is a defined report distribution requirement. During the interview with the incident investigation coordinator, he should determine if appropriate avenues such as worker safety committees, safety meetings, or posting incident reports were used to disseminate findings. The auditor should also verify that facility personnel and contractors are aware of recent incidents. This information should become a part of the process safety knowledge (see Chapter 4).

During the interview with the incident investigation coordinator, the auditor should determine if there is a policy regarding sharing incident reports with other plant locations. The auditor should also verify whether relevant reports have been distributed.

9.6.2 External Distribution

Leading companies in process safety often disseminate information on root causes of incidents to industry groups, particularly when similar manufacturing technology is involved. The auditor should determine if there is a company policy regarding external dissemination of incident causal information and if the policy has been followed.

9.7 Recommendation Implementation/Closure

The auditor should pay particular attention to this aspect of the incident investigation program. Closure on recommended actions should be obtained, regardless of whether they are accepted or not. The auditor should determine whether decisions about implementing recommendations have been documented. If there is a written procedure, the appropriate records should be obtained and compared with the recommendations resulting from the incident investigations. The auditor should determine if process modifications follow the management of change system (Chapter 6).

Effective and timely implementation of recommendations is important to prevent recurrence. During the interviews with the plant staff, the auditor should determine whether follow-up responsibility has been formally assigned for implementation of recommendations. The auditor should also determine if there is a system for follow-up and closure of action items, and if there are items that are outstanding beyond the required completion date.

In the process of auditing the incident investigation system, the auditor will have accessed the documentation needed to determine the extent of record keeping for incident investigations and follow-up. The auditor should verify that there is a written procedure on how the review is documented and that the appropriate follow up responsibility is assigned to a person with appropriate authority. Preferably, there should be a central depository for incident investigation records, rather than an individual's files. The auditor should evaluate the quality of the records in terms of completeness (i.e., properly filled-out incident reports, adequate investigation, follow-up, and status of recommendations, and such). The auditor should pay particular attention to whether the final status of all recommendations has been documented, regardless of whether they were implemented. The reasoning behind decisions that specify no action is required should also be documented.

9.8 Incident Analysis

As indicated at the beginning of this chapter, learning from mistakes should be the fundamental objective of incident investigation. Classification of findings by root causes provides the plant with valuable information about major contributors to incidents, and can assist facility management and corporate management in deciding where to place emphasis for improving process safety. Knowing whether incidents are resulting from facility failures (e.g, process, controls, and such) or deficiencies in management systems (e.g., procedures, training) is essential for proper management of process safety. The auditor should verify that incidents have been classified according to root causes.

Trend analysis provides plant management feedback on the effectiveness of the safety program implementation. It involves long-term tracking of incidents, grouped by root cause. Of equal importance is how this information is used. The auditor should verify that analysis of incident trends occurs and determine if appropriate process modifications or training program adjustments have been made as a result of the trend analysis.

9.9 Summary

Incident investigation is an important process safety management element to determine the root causes of incidents and thereby prevent similar incidents. The auditor should verify that there is a system to report and investigate all incidents. The system should include action to correct deficiencies and disseminate findings.

10

Human Factors

10.1 Overview

Human factors is the application of what is known about people to the design of technical systems and equipment to enhance safety and productivity through more efficient and effective human–system interactions. This is an emerging discipline for which there are few standards or guidelines. Human factors is an important component of any effective process safety management system. Often human factors are addressed informally rather than in a systematic manner throughout the life cycle of the process. Failure to address human factors in system design may result in decreased quality and productivity, low employee morale, increased absenteeism and turnover, and increased human errors. Ultimately it may lead to employee injuries and process safety incidents.

Although there is a human factors component in many of the other elements discussed in this book, it is critical that the auditing process explicitly address human factors because of the significant role that human error can play in process safety incidents. A process safety management system should consider the impacts that behavior, and physiological and psychological capabilities can have on the interface between the human and the working environment. Ergonomic and human factors checklists are available to assess the extent to which systems or equipment have been designed using human factors engineering practices. However, rather than concentrating on the results of failing to address human factors issues, this chapter focuses on auditing the process safety management system to ascertain how effective and comprehensive the system will be in implementing human factors principles and practices. Since there are few industry standards or guidelines in this area, the audit will generally be limited to company standards or practices.

When considering the interfaces between the operations or maintenance staffs and specific pieces of equipment, or the process as a whole, it is important to recognize that there are many human factors issues that may have an impact on the safety and performance of the system. Examples of questions that may be asked in a comprehensive human factors review are included in Table 10-1. The questions included can be used to help evaluate how effectively human factors have been incorporated into the process safety management system at a facility. Ineffective or incomplete application of human factors principles may result from a variety of reasons, including the nature of the task, the design of the equipment, the capabilities and limitations of personnel,

Examples of Items to Consider in a Human Factors Review

Organization and Policy Issues
- Have human factors engineering policies been established and communicated to employees by upper management?
- Are human factors support and expertise available within the organization?
- Is there a communication and follow-up mechanism to address human factors issues?
- Does the organization encourage employees to express human factors suggestions and concerns and allow them to contribute in the decision-making process?
- Is management willing to allocate time and resources to address human factors issues?
- Are human factors issues discussed at management meetings?
- Have formal procedures or policies been established that address the evaluation of new, modified, and/or existing processes/systems in terms of human factors principles?

Operator–Process Interface
- Do control and display layouts minimize the chance for operator error in terms of functional grouping, sequence of operations, and color (including use of colors appropriate for color blind individuals) etc.?
- Are there design standards that specify proper layout?
- Is information clear, concise, and readily accessible to the operator (including left-handed individuals)?
- Are the implications of various actions and their effects on the process clear to the operator?
- Are controls accessible, easy to reach, and operable?
- Are critical controls operated in the same manner (e.g., up–down; push–pull)?
- Is there adequate space to access system elements for normal operations and maintenance?
- Have upset conditions and emergency response been considered?

Task Design and Job Organization
- Have the operator's individual responsibilities been clearly defined? (See Chapter 3.)
- Have the psychological and physical demands of the job been considered for both routine and emergency operations?
- Have actions been taken to reduce the likelihood and impact of potential human errors?
- Have shift work and overtime schedules been designed to minimize operator fatigue/stress?

Work Place and Working Environment
- Have posture, movement, and accessibility been considered for both operations and maintenance activities?
- Have environmental conditions (e.g., noise, temperature, illumination, etc.) been considered?
- Have employees made modifications to existing systems that would indicate failure to apply human factors principles in the original design?

Training and Education (See Chapter 11.)
- Have people been assigned to jobs based on demonstration of required skills?
- Have process-related training requirements been defined based on job requirements?
- Have training modules and methods been developed?
- Does training include hands-on-exercises or simulations?
- Has the training been documented?
- Have employees received training in human factors?
- Are there training programs and support services to help employees with controlled substance use or abuse, or mental health problems?
- Have supervisors been trained in detecting the effects of substance abuse/stress on the performance of personnel?

Procedures (See Chapter 8.)
- Are procedures clear and complete, consistent in format and terminology, and compatible with the comprehension level of the user?
- Is there a system in place to ensure that procedures are periodically reviewed and updated based on process changes, the results of job/task analyses, or investigation of process incidents and near-misses?
- Have operators been involved in developing and/or reviewing operating procedures?
- Have critical operating procedures been clearly identified as such?
- Do procedures properly account for other activities for which the operator may be responsible at the same time?
- Have procedures been reviewed relative to the response time available to the operator to correct a problem?

or the characteristics of the operating environment. The auditor should focus on identifying the management system deficiencies that may have led to omission of human factor considerations rather than on identifying individual symptoms. The following sections provide guidance on identifying system deficiencies.

10.2 Organizational Issues

A company's organization and operating environment must support and promote the implementation of human factors principles and practices in day-to-day facility operations. Commitment and involvement are complementary and essential elements in an effective human factors program. Management may demonstrate commitment through such mechanisms as a well-communicated policy statement, a formal communication and feedback system, a written human factors program or procedure, and a willingness to provide adequate funding and authority to responsible parties.

In reviewing human factors programs, it is important to distinguish between those that have been simply designed to address injuries and illnesses, rather than those that deal with process design issues, such as controls, displays, and equipment layout. The auditor should determine what programs, policy statements, and feedback mechanisms exist and the extent to which they address process-related human factors issues. The effectiveness of the human factors programs and policies can be determined through the nature and frequency of their application and the knowledge of program content exhibited by facility personnel. The dates the various programs were initiated are also informative because a relatively new policy or program may not have been in place at the time the majority of the process was designed. In such cases, the auditor should query how the new program or policy has been integrated into existing operations (e.g., concurrent with a process hazard analysis).

A company's employees should be actively involved in the implementation of the human factors design program, as their first-hand experience will often provide early detection of human–hardware interface concerns or problems. The operating environment should encourage employees to express their suggestions and concerns, and allow them input into the decision-making process. Evidence of employee involvement might include representation on human factors or safety committees, participation in human factors design reviews or audits, or participation in the development of process operating and maintenance procedures or safety training programs. The auditor, for example, may review safety committee minutes, design review reports, job safety analysis reports, or safety inspection reports to determine whether operations and maintenance employees have been active participants in human factors programs, and at what stages. Involvement only in the identification of problems is not as desirable as involvement in the development and implementation of solutions. Interviews with personnel can help determine their actual level of involvement.

Knowledge and awareness of human factors among facility personnel, including management, operators, maintenance staff, and engineers, also indicate organizational support and commitment to the successful application of human factors principles to system design. The auditor should meet with personnel responsible for training and determine the extent to which facility training addresses human factors issues. The auditor should interview personnel to determine whether they have an understanding of human factors principles and their application to process safety management. He should also identify what human factors expertise is available onsite, from the corporate office, or from outside consultants or governmental agencies.

10.3 Design Considerations

During the design of a new process or system, the facility's policies or standards should promote the consideration of human factors issues during each stage of the development process. It is critical that human factors principles become an integral part of the design process, rather than an afterthought that is considered only when personnel are unable to adapt themselves to the system that has been devised. When process changes are made, systems that explicitly address human–hardware interface issues should be in place (see Chapter 6). Similarly, process safety reviews of existing processes (see Chapter 8) should include examination of human factors issues.

To verify whether and to what extent human factors is an element of the design review process for both new installations and modifications, the auditor should interview engineering and management personnel regarding the design standards that were followed for several recent projects and modifications. Considerations of space and equipment accessibility for maintenance should also be indicated in a good human factors program. Less formal involvement may be demonstrated by examining the attendance list and meeting notes for design review meetings. The use of mockups for panel layout, process models, or human factors reviews at the pilot plant level would be another indication of the incorporation of human factors principles in the initial design process.

The auditor should review any written procedures or checklists dealing with design of systems or ongoing reviews of system performance for incorporation of basic human factors principles (see Table 10-1). Design guidelines regarding line and equipment identification, labelling, type and placement of controls/alarms, and panel layout to ensure consistency between systems should also be available. The potential effects of environmental conditions and job-specific requirements (e.g., use of personal protective equipment, or access for maintenance) on the human–system interface should also be addressed in these design guidelines (see Section 10.6 for further details).

Once the auditor has a good understanding of how the facility's design review mechanism is expected to operate, he should review the operating areas to determine whether human factors principles have, in fact, been implemented. The purpose of this audit step is to identify any issues involving poor human factors design that may indicate a management system failure. The review should address the questions posed in Table 10-1, as well as specific findings from the interviews and document reviews.

10.4 Operating Culture

A commitment to human factors should also involve consideration of the effects of shift work, overtime, and other demands of the system on employees. The auditor should query personnel about workloads and overtime, and should verify that appropriate limits on overtime work have been specified and are followed.

The underlying way in which the system is organized, including task allocation, definitions of roles and responsibilities, and administrative practices, is important. Good human factors design includes more than just assuring that a monitor or control panel is located at the proper height. Stress and fatigue are often negative results of a failure to consider how people and the existing organizational structure will fit into the system that has been developed. The auditor should interview personnel and review design procedures to verify that psychological demands as well as the physical demands of the system or process have been considered. Problem areas may be identified by determining if there are particular activities that many operators or maintenance staff try to avoid, and why they are disliked.

Facility programs should promote the incorporation of human factors principles into the operating environment, and they should also specifically address factors that would have a negative impact on human performance. The use of drugs or alcohol in the workplace increases the likelihood of accidents or incidents resulting from human error. The auditor should interview facility personnel responsible for employee assistance programs and medical services to determine whether programs exist to educate and assist employees regarding drug or alcohol abuse. The auditor should also interview employees to determine the accessibility and effectiveness of such programs. In addition, he should determine whether the facility conducts periodic or pre-employment drug screenings for employees and contractors, and interview management and supervisory personnel to determine if they have received training in identifying substance abuse situations. Personnel required to take medication should report this condition to their supervisors.

10.5 Operating Procedures

The extent to which human factors has been incorporated into the operating environment of the facility can also be evaluated through a review of the facility's operating procedures. The auditor should review operating procedures to determine whether they are clear and complete, consistent in format and terminology, and compatible with the comprehension level of the operator. As mentioned previously, operator involvement in the development or review of procedures is desirable from a human factors standpoint. The auditor should determine, through interviews and a review of available documentation, whether operations personnel were involved in writing and/or reviewing operating procedures (see Chapter 4). In addition, the auditor should determine whether a formal mechanism or system is in place for periodic review and updating of the procedures from a human factors perspective, based on process changes, the results of job safety analyses, employee suggestions, or investigation of process incidents—particularly those that were influenced by human factors. The auditor should also interview operators to verify that the operating procedures are complete and well-matched to actual performance. He should also verify that qualified employees are filling in for staff shortages caused by vacations or illnesses.

The auditor should also review the usefulness of the procedures in helping operators to identify and respond to upset conditions. For critical operating procedures, he should address these questions:

- Do operators have adequate time to respond?
- Is there adequate verification that the job has been done properly (i.e., by cross-check by others or through instrumentation or controls)?
- Has parallel review of maintenance procedures been conducted?

10.6 Environmental Conditions

Illumination, noise, temperature, and other environmental conditions, such as chemical exposure and vibration, play an important role in the ability of humans to interact effectively with equipment or a system. The industrial environment may inhibit the operator's ability to perform. A single environmental factor may cause performance to deteriorate for a number of reasons. For example, excessive heat or humidity may reduce the work capacity of the operator. Similarly, excessive noise can mask audible alarms, interfere with essential communication, and lead to stress, or temporary hearing loss. It is therefore important that environmental conditions be addressed as part of an effective human factors program to control or remove the negative effects that these conditions may have on human–system performance.

10.6.1 Lighting

Lighting is an important element in the design of any system, as improper lighting levels may cause system elements to be seen incorrectly or not seen at all. Improperly designed lighting systems may result in eyestrain, muscle fatigue, headaches, or accidents involving slips, trips, or falls. Process incidents may also result from improper lighting levels which may cause an operator not to respond to visual displays or cues. The adequacy of lighting depends upon the type of lighting provided, its quality and quantity, the age of the worker, and the visual requirements of the task or system.

The audit of human factors should determine if a human factors review has been conducted to identify critical visual tasks within the system, and ensure that the lighting provided allows for those tasks to be performed safely and effectively. The auditor should determine whether facility personnel have performed lighting surveys, and what lighting standards have been used. The auditor should also interview operations personnel to determine whether there are problems with respect to illumination (e.g., levels, glare).

10.6.2 Noise

Excessive noise levels can interfere with communication, cause psychological stress, and alter one's performance level because of the annoyance or distraction.

The auditor should review available documentation to determine whether the facility has performed sound level surveys on existing equipment or specified noise levels of proposed systems. He should also note potential interference with critical communications or alarms. Where protective hearing equipment is required, the facility should provide alternate means of communication (e.g., visual alarms). The process safety management auditor should verify the existence and completeness of such reviews and surveys.

10.6.3 Temperature

The temperature in a work area can have a significant impact on how effectively an operator is able to perform. Hot conditions may result in a reduction in an operator's performance from fatigue or discomfort, or may even discourage the operator from performing that task. Outdoor work in cold weather can lead to decreased productivity and potential errors during manipulation tasks because of the need to wear gloves and thus a loss of dexterity.

The auditor should determine whether the facility has considered the effects of temperature in the design of new systems and procedures, or in process modifications, through interviews and a review of available documentation from recent projects. The auditor should also examine the extent to which engineering or administrative controls have been used to mitigate hazards caused by hot and cold environments. Examples of possible control measures include installing ventilation/heating systems, providing protective clothing, modifying work practices, or limiting the duration of exposure.

10.6.4 Other Environmental Conditions

In addition to noise, lighting, and temperature, other environmental conditions may have an impact on the effectiveness of the human–system interface. These environmental conditions may include vibration or chemical exposures (vapors, dusts, mists, or fumes). Vibration may lead to difficulties in reading critical display values or cause physical discomfort. Likewise, dust or chemical mists may obscure visual displays or cause eye or respiratory irritation.

As with the environmental conditions discussed previously, the auditor should determine the extent to which the system requirements have been identified in terms of the existing or potential environmental factors. The auditor should then conduct interviews and review available documentation to determine whether such environmental factors were addressed as part of the design and ongoing review process. She should also review with the operator adherence to and difficulties in carrying out certain procedures. Procedures and equipment should have provisions for operations and necessary maintenance in inclement weather, such as rain, snow, or high winds.

10.7 Process Control Issues

Process control-related design issues have become increasingly important as the use of computer-controlled process systems has become more widespread. Operators must be able to effectively interact with computer systems and the information displayed on video screens to ensure that the process runs safely and efficiently.

10.7.1 Display Design and Layout

Displays allow the machine or system to communicate with its user by providing cues, status reminders, and/or indications of conditions that require acknowledgment or a response by the operator. The design and installation of a display will affect the performance of the operator and ultimately have an impact on the effectiveness of the system as a whole. The displays should be located and spaced according to their function, relative importance, and relationship to their associated controls. The types of displays that are used depend upon the requirements of the task–system in terms of the type, amount, and required accuracy of the information to be communicated. Factors such as viewing distance, the number of displays, legibility, labelling, lighting, and glare should all be considered when selecting displays. Whatever displays are chosen, they should present information in a meaningful form that can be transmitted quickly and accurately to the operator. Displays should be configured in a way that facilitates monitoring, comparison, and sequencing so that normal, out-of-normal, and emergency conditions are readily apparent to the operator. In addition, displays should present no more information than is necessary to successfully operate the system.

The auditor should interview facility personnel responsible for the design and installation of computer displays to determine whether the design process includes consideration of human factors issues. In addition, the auditor should determine whether the operators (and maintenance staff, if appropriate) feel that the displays are clear, that they have been located properly, and that they provide the information needed for effective and efficient job performance. The treatment of operator concerns and suggestions can also be pursued as a means of verifying the feedback system.

10.7.2 Alarms

Alarms are an important means of notifying operators of changes in the process that may lead to process incidents. However, the alarm system can overload the operator with unneeded, supplementary information, or introduce unnecessary interruptions during emergency situations. It is therefore important that alarms which are associated with critical process safety parameters be distinguishable from those that provide information. If all signals are indicated in the same manner regardless of importance, these situations may become routine and the operator may choose not to acknowledge alarms. A priority system for alarms and/or a means to determine the initial alarm should exist for conditions where multiple alarms may occur.

The auditor should interview facility personnel to obtain an understanding of how decisions are made regarding alarm system design. The auditor should determine whether the alarm system design provides for the identification of critical process safety parameters; alarm criticality; and possible suppression of non-critical alarms during multiple alarm situations. The auditor should also verify that the operators understand the meaning and criticality of alarms.

Although an effective alarm system is important, the system should not serve as a substitute for careful monitoring of process parameters by the operator. Experienced, well-trained operators can identify potential problems by observing unusual trends and small fluctuations in process instrumentation. The system should not prohibit this monitoring, nor should this ongoing monitoring be capable of blocking out the alarm screen. The routine use of safety interlocks for process control (e.g., regularly using a high-level safety interlock as the normal way to discontinue tank filling) should not be allowed.

10.7.3 Match between Operations and Program

For computer displays of process systems to be effective in communicating information, the graphics and symbols must match the process in terms of flow and system components. Ideally, the display should provide some means of associating individual system components with the process as a whole. The auditor should interview operators regarding the effectiveness of computer displays in transmitting process information. If procedures are computerized, they should also match all of the steps performed by the operator. The auditor should also verify that there is a system in place at the facility for updating computer screens/graphics to reflect process or procedural modifications.

10.7.4 Monitoring Multiple Screens

The ability to monitor multiple screens effectively is essential to tracking process parameters. Screens should be arranged to facilitate viewing without requiring the operator to move about unnecessarily. For example, computer screens arranged in a U shape allow an operator to sit in the middle and view the screens by merely turning, while screens that have been arranged along a wall require operators to constantly walk the length of the control panel. If an operator is required to monitor multiple screens associated with different processes (e.g., batch operations), the processes must be timed or arranged so that the operator is not required to perform tasks simultaneously. This should be the case for both routine operations and for those situations where a batch may need to be held at temperature for longer than usual or some other variation from "normal." The auditor should interview facility operators to determine whether they are able to effectively monitor multiple screens and processes. In addition, he should interview system designers to determine if the requirements for monitoring multiple screens were considered during the layout of the displays and the development of computer programs.

10.8 Summary

The auditor should verify that there is a system in place to identify human factors issues in the design of new processes. For existing facilities, the auditor should verify that human factors issues are identified through process hazard analysis, incident investigation and other inspections.

11

Training and Performance

11.1 Overview

Training is a critical element of a process safety management system. Process safety management requires that people throughout an organization understand their roles, and have the knowledge and skills to perform in those roles. A training program helps everyone throughout an organization contribute to process safety.

The first part of this chapter presents a discussion of the issues that should be addressed by the auditor examining process safety training programs. The second part presents a general description of the types of training that should be included in an overall training program designed to enhance process safety.

11.2 Auditing of Training Programs

11.2.1 Needs Analysis

There are two types of process safety training that an organization should provide: general and specific. General training includes those subjects with which everyone employed at a facility should be familiar; specific training includes the subjects that are important to particular groups of employees.

The auditor should gain an understanding of the management system for training by asking questions, such as the following:

- How are process safety training objectives set?
- At what level of the organization is this done?
- How frequently is this assessment made, and by whom?
- At what level of the organization does this training take place?
- Who is assigned primary responsibility for the training?
- How is input provided from other groups to ensure that integration of a specific subject into the overall training program will be correct?
- What subjects are to be included in the training program?
- After development of the training materials, who has the responsibility and authority for validating the completeness and accuracy of the materials?
- Has the validation actually taken place?

- Additionally, has the need been met for a periodic review of the materials or lessons to ensure that they are still current and that the needed changes have been incorporated into them. (For example, one important input is from the management of change procedure.) Has it occurred?
- Has the training program been documented in a syllabus that is adequate? updated? used?
- Is there a provision for confirming that the material has been learned?
- Is there specification for the frequency of refresher training?

Once the overall process safety training needs have been determined and the implementation mechanisms established, it becomes necessary to determine the specific needs of a site, unit/department, function or individuals. The auditor should

Training Area / Plant Function	General Safety Training	General Process Safety Training	Process Specific Training	Task Specific for Operations	Task Specific for Maintenance/Contractor	Safe Work Practices	Process Safety Management Training	Emergency Response Training	Mandated Training
Site Management	X						X	X	X
Operations	X	X	X	X		X	X	X	X
Engineering	X	X	X				X	X	X
Laboratory	X	X						X	X
Maintenance Site Staff/Contract	X	X	X		X	X	X	X	X
Services Utilities	X	X	X	X		X		X	X
Material Handling	X	X	X					X	X
Waste Water	X	X	X	X				X	X

FIGURE 11–1. **Example Training Matrix for Site Organization**

TABLE 11-1

Examples of OSHA Required Training

Training Subject	Fed. Reg. Reference
Fork Truck Operation	29 CFR-1910.178
Lockout, Tagout, of Hazardous Sources	29 CFR-1910.147
Confined Space Entry (Pending)	29 CFR-1910.146
Respirator Training	29 CFR-1910.134
Overhead Cranes	29 CFR-1910.179
Emergency Response	29 CFR-1910.120
Employee Emergency Plans and Fire Prevention Plans	29 CFR-1910.38
Hazardous Communication Standard	29 CFR-1910.1200
Hearing Conservation	29 CFR-1910.95
Process Safety Management of Highly Hazardous Chemicals	29 CFR-1910.119

verify who determines the needs, the methodology used, and how frequently this needs assessment is updated to reflect changes in process technology, equipment, tools, regulatory requirements, and personnel (see Figure 11-1 and Table 11-1).

Most process safety management elements require training. Generally this training is coordinated centrally, but some training may be addressed within specific elements. One key element that needs to interface closely with the training element is management of change. The auditor should have a clear understanding of who has responsibility for various training needs, particularly those specific to other elements.

Most process safety training will cover new employee(s) and employee(s) transferred between units or sites. The auditor should review the process for determining the training requirements of a transferred employee or one who fills a position temporarily such as when someone is ill or on vacation.

Training that contractors must receive before they can work on site should also be identified. The contractor must be provided with information on the special hazards that relate to his work area, but these requirements may differ among different types of contractors. For example, a contractor who performs maintenance will often require different information from a contractor who performs on-site construction. The auditor should confirm that a determination of what training must be provided to contractor representatives has been made.

The auditor should also review the system in place by which the company confirms that contract employees have been trained by the contractor. This company system might involve review of the contractor's training records, and interviews with the contractor employees.

The personnel in charge of the training program should be responsible for measuring its effectiveness. The auditor should not be charged with determining the effectiveness of the programs, but only monitoring their quality and verifying that they are conducted on a regular and systematic basis.

11.2.2 Program Content and Presentation

The design of a training program should be based on the assessment of needs. Training programs in any process plant cover a wide range of subjects and plant functions. However, to be effective training should be integrated to ensure consistency among subjects and plant functions. Examples of some specific training areas and programs, and how they fit into a process safety management system are discussed below.

The auditor should confirm the existence of a clear specification or guideline stipulating who is to present the course (i.e., a specific individual or a group of people from a particular classification or background). The auditor should also look at the qualifications of the instructors of completed courses, and their training to prepare themselves for the instructional task.

The auditor should verify that training modules address the needs identified in the needs assessment. The actual training course material might consist of a lesson plan, an outline, a course textbook or workbook, or a videotape. The course documentation should address, and should allow the auditor to understand, the essence of the training program.

GENERAL PROCESS SAFETY TRAINING

General process safety training should cover the plant safety rules; emergency alarm signals such as alerts, evacuation, and all-clear; and smoking and no smoking areas in the plant. Other training might cover emergency reporting, summoning aid for an injured person, or reporting a fire or the spill of a hazardous substance.

PROCESS-SPECIFIC TRAINING

Training subjects covered under the process-specific category should deal with basic process chemistry, process design, and critical parameters that affect the reactions and productivity of the unit. These subjects are general overviews of the process.

Important process safety considerations of the raw materials, process streams, and products relating to flammability, explosivity, and toxicity should all be covered in the training curriculum. This training should be presented to personnel involved in the process, including operators, maintenance personnel and contractors.

TASK-SPECIFIC TRAINING FOR OPERATING PERSONNEL

Task-specific training generally starts with the presentation of basic principles and then becomes specific for the process under consideration. For instance, general training could be offered in the common unit operations found on the site such as mixing, distillation, reaction, separation, heat transfer, and pump and compressor operation. An additional area of general training might cover the principles of process control. The concepts of the basic control loop, feedback, and measurement of basic process parameters such as flow, pressure, temperature, and level, are common to all processes, and thus an understanding of these concepts is needed by all operators.

Once this general information has been taught, trainees will have a basis for the specific unit training.

Specific unit training would then concentrate on the details of the units that comprise the process. Trainees would first learn how a specific unit fits in the overall process, and then how it must be set up to put both the process equipment and the associated instrumentation into operation. Next the normal sequence of operation would be covered, using the standard operating procedures (SOP) used to run the process (see Chapter 4). Some possible training areas that might be covered in these SOPs include:

- Established safe operating limits
- Consequences of operations outside these limits
- Actions to take to keep from exceeding safe operating limits
- Equipment setup
- Startup procedures
- Normal operation
- Possible process hold-points
- How to restart operation from these hold-points
- Normal process shutdown
- Emergency procedures
- Equipment troubleshooting

TRAINING FOR MAINTENANCE/CONTRACTOR PERSONNEL

Specific training for personnel performing maintenance in process areas should include the general process safety training and relevant aspects of process specific training. The training should then focus on standard maintenance operations, such as working in specific hazardous locations and the maintenance tools and techniques unique to that craft. Specific training in preventive maintenance procedures and techniques should also be included (see Chapter 7). If maintenance/engineering personnel are required to conduct quality assurance tests and inspections during the fabrication and construction of new facilities, they should receive appropriate training in these areas as well. Facility staff should provide contractors with process hazard and site specific information necessary for them to train their employees. The auditor should verify that there is a system in place to assure that contractors are trained.

TRAINING FOR THE OPERATIONS–MAINTENANCE INTERFACE

The operations function generally prepares equipment for maintenance. At this point, an interface between operators and maintenance personnel exists. Therefore, both groups should be trained in the requirement of work authorization procedures and safe work practices. Operating personnel need to know how to prepare the line or equipment for maintenance, what tests have to be performed to ensure it is safe for maintenance work, and how to conduct these tests.

Maintenance personnel must know what state the equipment should be in at the end of the job before returning it to the operations. Operations should verify that equipment is safe to operate before starting up.

PROCESS SAFETY MANAGEMENT TRAINING

In addition to information discussed earlier, training in Management of Change procedures should be provided (see Chapter 6) for some personnel. Process safety training should also include techniques for conducting process hazards analysis and risk assessment for those who will participate in such analyses (see Chapter 8). Training in the techniques of accident/incident investigation is also necessary (see Chapter 9) for some individuals. Specific training in emergency response is addressed in Chapter 12.

11.2.3 Training Frequency

Training programs should be presented frequently enough to maintain skills. Some of the regulatory agency-mandated programs listed in Table 11-1 require retraining at set intervals; other training must be given when changes have taken place as part of the overall management of change program. There may be some operations training that should be given each time certain infrequent tasks are conducted, such as preparing a unit for turnaround or startup following a turnaround. The auditor should confirm that training frequency requirements are specified and are being met.

11.2.4 Training Records

The auditor should verify that training records include a description of the training course, the date it was presented, the instructor, and the names of the attendees. Attendance records should be compared with employee rosters, or training schedules, to confirm that all employees are being trained within a specified period. It is important to verify that the system provides for make-up sessions for personnel who were ill or on vacation when training was conducted.

The training records may be kept in a central location or in the individual departments. It might consist of a separate paper file on each of the courses, or a computerized data base from which lists of courses and attendees may be printed out.

11.2.5 Training Program Effectiveness

Mechanisms that can be used to determine the effectiveness of a training program are many and varied. There are both short and long-term indicators of effectiveness. Short-term indicators are:

- Written and oral testing
- Demonstration in the field

Long-term indicators are

- Random spot-check testing
- Incident reports
- Log sheets and log books

The auditor should interview employees who have had different types of training to obtain their view of its effectiveness.

If the facility has a certification requirement under which certain types of employees must qualify prior to assignment, the auditor should confirm that employees in those positions (e.g., welders) have met the requirements.

11.3 Summary

Training is a key element of process safety management. This chapter discussed how to audit specific training related to operating and maintaining a process. Many other elements of process safety management require specific training to be effective, as discussed in other chapters. The auditor should verify that a thorough training needs assessment has been conducted and that the training program is well designed and implemented.

12

Emergency Response Planning

12.1 Overview

Emergency response planning embraces a wide range of activities aimed at mitigation or control measures for process upsets, fires, explosions, spills, chemical releases, and other sudden, unplanned events that might result in damage or loss. Emergency response planning objectives might include measures necessary to prevent or limit losses in one or more of the following areas:

- Acute health effects to workers, emergency responders, and the public
- Environmental damage
- Property, equipment, or product damage
- Production loss
- Loss of good will and public trust, and
- Third-party liabilities

12.2 Needs Analysis

The auditor should verify that the system for developing emergency response plans has identified potential hazards of the facility's processes. The types of hazards identified in process hazards analysis and other risk assessments (as described in Chapter 8) should be addressed in the emergency response planning process. Historical incident data provide another source of information for developing response plans. Other factors that should be considered include the nature of the process hazards, the potential consequences, the loss control objectives of the facility, the proximity of the facility to public areas and environmentally sensitive areas and the availability of internal and external resources such as facility staff, emergency response agencies, water supplies, and equipment. The auditor should verify that systems are in place to address these issues, and that there is a system to review and update emergency response plans when facility modifications change the nature of the hazards.

Emergency response planning is affected by a host of regulatory requirements. An auditor of emergency response planning should verify that a mechanism is in place to identify and address applicable regulations.

TABLE 12-1

Examples of U.S. Emergency Response Planning Regulations

- OSHA 29 CFR 1910.3(a)—Employee Emergency Plans and Fire Prevention Plans
- OSHA 29 CFR 1910.156—Fire Brigades
- OSHA 29 CFR 1910.120 (q)—Hazardous Waste Operations and Emergency Response
- Resource Conservation and Recovery Act (RCRA)
- Superfund Reauthorization and Amendments (SARA Title III)
- Clean Air Act (CAA)
- National Pollutant Discharge Elimination System (NPDES)

Emergency response plans are guided by national, state, and local regulations. Examples of U.S. Federal regulations are noted in Table 12-1.

12.3 Emergency Response Plan Content

Most facilities have prepared a variety of emergency response plans and procedures for events such as spills, fires, explosions, chemical release, power outages, and operational upsets. It is important that the auditor clearly identify the range of planning that has been performed by the facility.

As a minimum, the auditor should ensure that emergency response planning addresses the safe control of processes in emergency conditions, and instructions and training of employees and contractors to minimize risks to life, the environment, and property. The minimum provisions should include:

- Alarm and notification
- Emergency evacuation and/or shelter
- Spill containment and control
- Loss of critical power and utilities
- First aid medical care, and
- Response procedures to fires, explosions, and chemical releases

The auditor should identify that the following basic elements of emergency response planning are in place:

- Identifying hazards
- Education and training
- Planning
- Testing and maintenance
- Conducting drills/exercises, and
- Critique

Figure 12-1 illustrates the relationship of elements in emergency response planning.

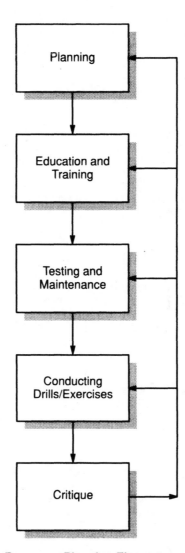

FIGURE 12–1. **Emergency Response Planning Elements**

In reviewing emergency response planning documents, the auditor should determine whether a management system is in place to respond to a major emergency, and whether planning details, such as strategy, procedures, supplies, resources, and organization have been defined to facilitate training and to maintain the plan. A complete emergency response planning system should include mechanisms for accomplishing each of the emergency response planning elements.

12.4 Auditing Emergency Response Planning

In preparing for the audit, the auditor should identify the key individuals with responsibility for emergency response planning. Typical documentation which should be available to the auditor include:

- Facility description, including organization and staffing
- Emergency response plans
- Plot plans
- Facility or company policy regarding emergency response planning
- Description of emergency systems and equipment
- Emergency organization plan
- Description of process hazards
- Description of external resources and support organizations
- Regulations applicable to the facility.

Some or all of these items may be contained in a comprehensive emergency response plan, or they may be maintained in separate documents. The information will be useful in understanding the scope and depth of the emergency response planning effort and the needs of the facility.

The audit of this element should include the following steps:

- Review of documentation of the facility organization and staffing as described above
- Review of emergency action plans and procedures (including external planning efforts)
- Interviews with department managers to understand management systems in place
- Sample records for compliance with procedures and completeness
- Interviews with emergency response personnel, other employees, and contractors to verify that training requirements have been met
- Sample emergency systems and equipment specified in the emergency response plans, and
- Review of critique on conducted drills/exercises

12.5 Emergency Management System

The auditor should verify that the facility has in place an incident command system to manage emergency responses. Specifically, the auditor should examine emergency plans to determine if the following characteristics can be identified and that systems are in place to specifically test and exercise each:

- Integrated communications
- Comprehensive resource management

- Predesignated command facilities
- Consolidated action plans
- Manageable span of control, and
- Modular organization

Appendix A provides a more detailed description of characteristics that may be included in an incident command system.

12.6 Documentation of Resources, Organizations, and Procedures

As illustrated in Table 12-2, the list of typical considerations of an emergency response plan can be very lengthy. An audit of these components should include:

- Identification of components in place
- Determination of a sampling strategy for each group of components
- Verification, through sampling, that the components described in the plan are in place, maintained and functioning

The auditor should verify that the required tests and inspections for maintenance of emergency response equipment have been conducted. Some equipment may be maintained under the general plant preventive maintenance program (see Chapter 7).

TABLE 12-2
Typical Emergency Response Plan Considerations

A. Company Policy and Plan Objectives
B. Facility Planning Basis
C. Emergency Response Organization-Structure and Duties
D. Detection, Alarm, and Notification Procedures
E. Emergency Communication Systems
F. On-site Evacuation and Security
G. Emergency Facility Shutdown Procedures
H. Medical Emergency Procedures
I. On-site Emergency Response Teams
J. Personal Protection of Response Teams
K. Fire Response Procedures
'.. Spill Containment and Cleanup Procedures
M. Environmental and Spill Monitoring
N. Public Relations in Emergencies
O. Application of Plan to Natural Hazards
P. Off-site Post-incident Recovery
Q. Off-site Sources of Assistance
R. Resource Listings-Supplies and Supplemental Services
S. Hazardous Material Data Sources

12.7 Training and Records

Training requirements for emergency response personnel, other employees, and contractors should be documented. These records should include the types of training required, training frequency, and a specific list of employees who are required to be trained under each category. The auditor should sample training records to verify that training has been conducted according to the documented requirements (Chapter 11).

12.8 Emergency Response Plan Effectiveness

A facility with a well-executed process safety management system will not have the number or size of incidents that would keep emergency responders skilled in the requirements of their assigned duties. An emergency plan, to be effective, must be periodically tested by emergency management exercises and drills.

Emergency management exercises are essential to assure that education and training are transformed into an effective response capability. Exercises can consist of a wide range of activities including drills, command center exercises, and full deployment of available resources. The complexity and scope of exercises appropriate for a facility will depend upon the nature of the risks, size of the facility, the legal authority for action by the facility, and the current level of preparedness. A facility with a newly written emergency plan would be ill-advised to test its plan with a full deployment, until that capability has been effectively demonstrated by a series of smaller exercises. Emergency exercises, like any activity, involve some level of risk, and they must be carefully planned and controlled. The auditor should verify that exercise schedules are being met, the exercise results documented, and changes made where appropriate.

12.9 Summary

An audit of emergency response planning should address the systems for identifying hazards and risks which the plan must address. The auditor should verify that an emergency response plan and its components are in place, maintained and functioning.

APPENDIX A
Characteristics of an Incident Command System

Incident command may call for critical decision-making within a very short time frame. Confusion over authority and responsibility must be minimized to reduce the likelihood that decisions will be delayed. An Incident Command System should clearly establish both the authority and the responsibility for each part of the emergency organization. Authority and responsibility of facility management, the company, and any outside response groups are important to identify in as much detail as possible. Jurisdictional disputes, whether within a facility or between emergency response agencies, need to be resolved before an emergency occurs.

A. Integrated Communications

Communications during an emergency can become a complex problem as the size of the incident increases. Communications considerations include:

- *Employee Notification:* The location and nature of the emergency need to be communicated, along with any instructional messages, to assure that employees know what actions are necessary. Prepared and consistent messages are usually more easily understood by employees than messages formulated during a crisis.

- *Communication among Emergency Responders:* Communication methods, from hand signals to radio messages, need to follow established protocols to be effective. Consistent labeling of equipment, streets, buildings, and other features of a facility need to be clearly established before an emergency. The timing, contents, and paths of communications within emergency response groups need to be known, planned, and practiced before an emergency occurs to prevent garbled communications, overloading of communications paths, and security breaches. Information needs for each response group should be identified in the planning stages to assure that the necessary paths for information flow exist or are established as needed.

- *Equipment Considerations:* Communications equipment considerations include compatibility of equipment, number and distribution of devices, and reliability. Plant intercom systems, process unit alarms, general alarm systems, radios, pagers, and other available communications devices should be integrated into the plan to support the incident command system. Portable cellular telephones are a popular emergency communications tool. However, it should be recognized that communications circuits can become overloaded in a regional or catastrophic event.

- *Public Communications:* Communications with the general public should be carefully planned before an incident occurs. Inappropriate or inaccurate information may do more damage than good if it incites panic or civil disobedience. Public communications involving the need for emergency actions by the public should be carefully planned with local officials before an emergency occurs to assure that the authority to communicate, the path of the communication, and the actions recommended are understood by all.

B. Comprehensive Resource Management

Resources include manpower and the hardware, foams, dispersants, booms, hoses, safety equipment, protective clothing, backup manpower, heavy equipment, boats, communications equipment, technical resources, instrumentation for environmental monitoring, specialty services, food, clothing, temporary housing, and the many other items that may be needed in an emergency. Resources for an emergency response should be identified and listed in an emergency response plan. Resources can come from internal and external sources. Some of the sources that may be considered are

- Company resources at other parts of the organization
- Public fire departments
- Police
- Spill cooperatives
- Mutual aid organizations
- Company response teams
- Hospitals
- Disaster relief agencies (e.g., Red Cross)
- Local emergency planning committees
- Hazardous materials advisory councils
- Civil defense organization
- Port authorities
- Coast Guard
- National Guard
- Volunteer agencies

Alternative sources for critical services and supplies should also be identified.

C. Predesignated Command Facilities

Command facilities may range from mobile field units to dedicated buildings, depending on the needs of the facility. In siting the command facility, consideration should be given to the safety and security of the command personnel. The design of the facility might consider the intended occupancy, communications needs, utilities, facilities for extended stays, information resources, provisions for separate meetings or conversation. Alternate command facilities may be needed on the possibility that the primary facility could be unusable because of wind direction, damage, or any other unforeseen event.

D. Consolidated Action Plans

An important goal of emergency response planning is to integrate the response efforts of each segment of the incident command system and any outside resources that are available. Action plans should be designed to take maximum advantage of fixed protection systems, skills of specialized response teams, and the support of the entire incident command system. Action plans should be developed for anticipated events and practiced in emergency drills to demonstrate their effectiveness. Safety of emergency responders should be specifically addressed in the action plans. For management and support positions of the incident command systems action plans should be developed to provide for quick deployment of resources, notification of outside and company response groups, organization of relief and support efforts, and quick reaction to anticipated changes in the status of the emergency.

E. Manageable Span of Control

Emergencies are often stressful events that demand decision-making under the most adverse conditions. Information is often received faster than it can be verified, analyzed, and acted upon. The organizational structure of an incident command system should limit the number of people or functions assigned to an individual; five to seven direct reports should be considered maximum. In an emergency, a vertical organization is desirable because it provides closer supervision and control.

F. Modular Organization

The appropriate size and complexity of an emergency response organization is a matter of need. Both in the planning and in the activation of an emergency, the efficient deployment of resources requires that those resources most needed be closest to the central command element of the plan. To accomplish an efficient and streamlined organization, resources are often organized and deployed in smaller units. Modules can be organized for both particular skills and for functional characteristics. A full deployment of fire fighting resources, for instance, might not involve deployment of spill containment resources if no spill hazard is imminent. A modular organization will also allow special skills, such as spill response teams, triage units, and communications units to be more easily identified.

Bibliography

Adams, E. E., *Accident Investigation Procedure - Some Guidelines for Classification,* Professional Safety, August 1985.

American Petroleum Institute (API), *Recommended Rules for the Design and Construction of Large, Welded, Low-Pressure Storage Tanks,* Recommended Practice 620.

American Society of Mechanical Engineers (ASME), *Rules for Construction of Pressure Vessels,* Section VIII, Divisions 1 and 2, Boiler and Pressure Vessel Code.

American Society of Mechanical Engineers (ASME/ANSI), *Chemical Plant and Petroleum Refinery Piping, ANSI/ASME B31.3.*

American Petroleum Institute (API), *Recommended Practice 750 — Guidelines for Management of Process Hazards,* Washington DC, 1988.

Burk, A.F., *Process Safety Management,* paper 77a presented at the Loss Prevention Symposium, AIChE 1988 Spring National Meeting, New Orleans, March 6-10, 1988.

Burk, A. F., *What-If/Checklist - A Powerful Process Hazards Review Technique,* presented at the American Institute of Chemical Engineers Summer National Meeting, August 1991.

Center for Chemical Process Safety (CCPS), *A Challenge to Commitment,* American Institute of Chemical Engineers, New York 1987.

Center for Chemical Process Safety (CCPS), *Guidelines for Safe Storage and Handling of High Toxic Hazard Materials,* American Institute of Chemical Engineers, New York, 1987.

Center for Chemical Process Safety (CCPS), *Guidelines for Technical Management of Chemical Process Safety,* American Institute of Chemical Engineers, New York 1989.

Center for Chemical Process Safety (CCPS), *Guidelines for Chemical Process Quantitative Risk Assessment,* American Institute of Chemical Engineers, New York, 1989.

Center for Chemical Process Safety (CCPS), *Guidelines for Hazard Evaluation Procedures, Second Edition with Worked Examples,* American Institute of Chemical Engineers, New York 1992.

Center for Chemical Process Safety (CCPS), *Plant Guidelines for Technical Management of Chemical Process Safety,* American Institute of Chemical Engineers, New York 1992.

Center for Chemical Process Safety (CCPS), *Guidelines for Investigating Chemical Process Incidents,* American Institute of Chemical Engineers, New York, 1992.

Chemical Manufacturers Association (CMA), *Responsible Care A Resource Guide for the Process Safety Code of Management Practices,* Washington, DC, October 1990.

Chemical Manufacturers Association (CMA), *A Manager's Guide to Reducing Human Errors,* Washington, DC, July 1990.

Chemical Manufacturers Association (CMA), *Process Safety Management,* Washington, DC, May 1985.

Cox, R.A., *An Overview of Hazard Analysis,* International Symposium on Preventing Major Chemical Accidents, February 1987.

Environmental Protection Agency (EPA), *Why Accidents Occur: Insights from the Accidental Release Information Program,* Chemical Accident Prevention Bulletin, OSWER-89-008.1, Series 8, No. 1, July 1989.

131

Ferry, T.S, *Accident Investigation and Analysis, A Dozen Steps for the Safety Professional,* Professional Safety, January 1981.

Greeno, J.L., et al., *Environmental Auditing—Fundamentals and Techniques, Second Edition,* Arthur D. Little, Inc., Cambridge, MA, 1987.

Gressel, M. G., and Gideon A., *An Overview of Process Hazard Evaluation Techniques,* American Industrial Hygiene Association J. Vol. 52, April 1991, pp 158-163.

Jacobs, H.C., *Improve Process Safety Reviews,* Hydrocarbon Processing, Vol 68, July 1989, pp 66-72.

John Gray Institute (JGI), *Managing Workplace Safety and Health: The Case of Contract Labor in the U.S. Petrochemical Industry,* July 1991.

Krivan, S. P., *Avoiding Catastrophic Loss: Technical Safety Audit and Process Safety Review,* Professional Safety, February 1986, pp 21-26.

Lawley, H.G., *Operability Studies and Hazard Analysis,* Chemical Engineering Progress, Vol. 70, No. 4, April 1974, pp. 45-56.

Lesins, V. and Moritz, J. J., *Develop Realistic Safety Procedures for Pilot Plants.* Chemical Engineering Progress, January 1991.

Munson, R.E., *Process Hazards Management in DuPont,* Plant/Operations Progress, Vol. 4, No. 1, January 1985, pp 13-16.

National Safety Council (NSC), *Accident Investigation...A New Approach,* Chicago, IL, 1983.

Occupational Safety and Health Administration (OSHA) 29CFR 1910.119, *Process Safety Management of Highly Hazardous Chemicals, Explosives and Blasting Agents; Final Rule,* February 24, 1992.

Olsen, R. E., *Incident Investigation, Reporting and Analysis,* presented at the CCPS-NAM Workshop on Process Safety Management for Smaller Companies, Atlantic City, New Jersey, June 7, 1991.

Ozog, H., and Stickles, R.P., *Process Hazard Management Documents, Practices Compared,* Oil and Gas Journal, January 28, 1991.

Ozog, H. , and Bendixen, L.M., *Hazard Identification and Quantification,* Chem. Eng. Prog., Vol. 83, No. 4, April 1987, pp. 55-64.

Shafagi, A., and Gibson S.B. , *Hazard and Operability Study—A Flexible Technique for Process System Safety and Reliability Analysis,* presented at the American Chemical Society Symposium Series 274 Conference titled Chemical Process Hazard Review, 1985, pp. 33-39.

State of California (California), *State of California Guidance for the Preparation of Risk Management and Prevention Program,* California Office of Energy Services, November 1989.

Stickles, R.P., Ozog, H., Long, M.H., *Facility Major Risk Survey,* presented at American Institute of Chemical Engineers Spring National Meeting, March 18-22, 1990.

U.S. Environmental Protection Agency (USEPA), *Review of Emergency Systems,* Report to Congress, Section 305(b), Title 111, Superfund Amendments and Reauthorization Act of 1986, U.S. Governing Printing Office, Washington, DC, 1988.

Wells, G.L., *Safety Reviews and Plant Design,* Hydrocarbon Processing, Vol. 60, No.1, January 1981.

Index